WORKSHOP PHYSICS® ACTIVITY GUIDE
THIRD EDITION

Activity-Based Learning

THE CORE VOLUME
WITH
MODULE 4: ELECTRICITY AND MAGNETISM

Electrostatics, DC Circuits,
Electronics, and Magnetism
(Units 19-27)

**PRISCILLA W. LAWS,
DAVID P. JACKSON AND BRETT J. PEARSON**

WILEY

SENIOR DIRECTOR	Michelle Renda
EXECUTIVE EDITOR	John LaVacca III
ASSISTANT EDITOR	Hannah Larkin
SENIOR MANAGING EDITOR	Judy Howarth
PRODUCTION EDITOR	Mahalakshmi Babu
COVER PHOTO CREDIT	© David Jackson and Brett Pearson

This book was set in 10.5/12.5 Aldus LT Std Roman by Straive™.

Founded in 1807, John Wiley & Sons, Inc. has been a valued source of knowledge and understanding for more than 200 years, helping people around the world meet their needs and fulfill their aspirations. Our company is built on a foundation of principles that include responsibility to the communities we serve and where we live and work. In 2008, we launched a Corporate Citizenship Initiative, a global effort to address the environmental, social, economic, and ethical challenges we face in our business. Among the issues we are addressing are carbon impact, paper specifications and procurement, ethical conduct within our business and among our vendors, and community and charitable support. For more information, please visit our website: www.wiley.com/go/citizenship.

ISBN: 978-1-119-85661-0

Library of Congress Cataloging-in-Publication Data

Names: Laws, Priscilla W., author.
Title: Workshop physics activity guide / Priscilla W. Laws, David P.
    Jackson and Brett J. Pearson.
Description: Third edition. | Hoboken, NJ : Wiley, [2023]- | Includes
    index. | Contents: Core volume with module 1. Mechanics I — Core volume
    with module 2. Mechanics II — Core volume with module 3. Heat,
    temperature, and nuclear radiation — Core volume with module 4.
    Electricity and magnetism.
Identifiers: LCCN 2023016539 | ISBN 9781119856528 (module 1 ; paperback) |
    ISBN 9781119856504 (module 1 ; epub) | ISBN 9781119856559 (module 2 ;
    paperback) | ISBN 9781119856535 (module 2 ; epub) | ISBN 9781119856580
    (module 3 ; paperback) | ISBN 9781119856566 (module 3 ; epub) | ISBN
    9781394227877 (module 3 ; pdf)
Subjects: LCSH: Physics—Experiments.
Classification: LCC QC33 .L39 2023 | DDC 530.078—dc23/eng/20231010
LC record available at https://lccn.loc.gov/2023016539

The inside back cover will contain printing identification and country of origin if omitted from this page. In addition, if the ISBN on the back cover differs from the ISBN on this page, the one on the back cover is correct.

SKY10078860_070324

CONTENTS

PREFACE

The principle of science, the definition almost, is the following: The test of all knowledge is experiment....But what is the source of knowledge? Where do the laws that are to be tested come from? Experiment, itself, helps to produce these laws, in the sense that it gives us hints. But also needed is imagination to create from these hints the great generalizations—to guess at the wonderful, simple, but very strange patterns beneath them all, and then to experiment to check again whether we have made the right guess.

—Richard Feynman, *The Feynman Lectures on Physics*

This is the third edition of the Activity Guide developed as part of the Workshop Physics Project. Although this Guide contains text material and experiments, it is neither a textbook nor a laboratory manual. It is a student workbook designed to serve as the foundation for a two-semester, calculus-based introductory physics course sequence that is student-centered and focuses on hands-on learning. The activities have been designed using the outcomes of physics education research and honed through years of classroom testing at Dickinson College. The Guide consists of 28 units that interweave written descriptions with activities that involve predictions, qualitative observations, explanations, equation derivations, mathematical modeling, quantitative experimentation, and problem solving. Throughout these units, students make use of a flexible set of computer-based data-acquisition tools to record, display, and analyze data, as well as to develop mathematical models of various physical phenomena.

The Activity Guide represents a philosophical and pedagogical departure from traditional physics instruction. Students who study science in lecture-based courses are presented with definitions and theoretical principles. They are then asked to apply this knowledge to the solution of textbook problems and the completion of equation-verification experiments. A major objective of Workshop Physics is to help students understand the basis of knowledge in physics as a subtle interplay between observations, experiments, definitions, mathematical descriptions, and the construction of theoretical models. Instead of spending time in lectures, students in Workshop Physics make predictions and observations, do guided derivations, and learn to use computer tools to develop mathematical models of phenomena they are observing firsthand.

There are several reasons for emphasizing the processes of scientific investigation and the development of investigative skills. First, the majority of students enrolled in introductory physics courses at the high school and college levels do not have sufficient concrete experience with physical phenomena to fully comprehend the theories and mathematical derivations presented in lectures. Second, the current body of physics knowledge is truly overwhelming, and the traditional lecture method often results in trying to cover too many topics, which can lead to rote memorization on the part of the students. We believe that the only viable approach is to help students master the fundamentals and develop strategies for learning other topics independently. Finally, through many years of student evaluations we have found that most students prefer an active method of learning because it provides an environment in which asking questions and trying to clarify their understanding is encouraged.

USING THE ACTIVITY GUIDE IN DIFFERENT INSTRUCTIONAL SETTINGS

This Activity Guide was originally designed to be used in relatively small classes in an instructional setting that combines laboratory and computer activities with discussion. Students work in collaborative groups of 2, 3, or 4 depending on the nature of each activity. Most of the activities in this Guide were tested and refined over an eight-year period at Dickinson College in a workshop environment, where sections of up to 24 students met three times a week for 2-hour class sessions. Over the years, this workshop style has been emulated at other institutions and expanded to class sizes of 50 students or more. In addition, many instructors have successfully adapted the activities for use in algebra-based courses, including courses at the high-school level. With careful planning, this Guide can even be adapted for use in large university settings. For example, many activities can be refashioned as a series of interactive lecture-demonstrations, and most can be used with little modification in smaller tutorial sessions. Thus, we believe the Activity Guide can be used effectively in almost any educational setting, particularly when in the hands of a motivated instructor.

TOPICS COVERED

To accommodate the time required for active, hands-on learning, there are slightly fewer topics discussed in this Guide compared to a traditional lecture course. We have retained topics that are most often covered in conventional courses: Newtonian mechanics, thermodynamics, and electricity and magnetism (including circuits), and eliminated topics we felt were less essential (or often covered in later courses): fluids, waves, and optics.

Although the coverage of traditional content has been reduced somewhat, we include some topics that are not typically treated in conventional introductory physics courses: Unit 15 on Oscillations, Determinism, and Chaos; Unit 25 on Electronics; and Unit 28 on Radioactivity and Radon Monitoring.

This Activity Guide is distributed in four different modules:

Module 1: Mechanics I
Kinematics and Newtonian Dynamics (Units 1–7)
ISBN: 9781119856504 (epub); 9781119856528 (print)

Module 2: Mechanics II
Momentum, Energy, Rotational and Harmonic Motion, and Chaos (Units 8–15)
ISBN: 9781119856535 (epub); 9781119856559 (print)

Module 3: Heat, Temperature, and Nuclear Radiation
Thermodynamics, Kinetic Theory, Heat Engines, Nuclear Decay, and Radon Monitoring
(Units 16–18 and 28)
ISBN: 9781119856566 (epub); 9781119856580 (print)

Module 4: Electricity and Magnetism
Electrostatics, DC Circuits, Electronics, and Magnetism (Units 19–27)
ISBN: 9781119856597 (epub); 9781119856610 (print)

The Complete Set: Modules 1–4 (Units 1–28)
ISBN: 9781119856375 (epub); 9781119856498 (print)

A typical year-long introductory physics course would use Modules 1 and 2 in the first semester and Modules 3 and 4 in the second semester, with a pace of approximately one unit per week. As with most introductory physics courses, the concepts build on each other, so we recommend covering units in numerical order, at least up through Unit 11. At that point, it's possible to skip around a little more easily (though most of the units in Module 4 should probably be covered in order).

COMPUTER-BASED DATA-ACQUISITION SYSTEMS

Computer-based data-acquisition systems are used extensively throughout the Activity Guide for the collection, analysis, and real-time display of data. Indeed, this is one feature of Workshop Physics that is particularly powerful. The use of real-world data and graphical representations provide an immediate picture of how a physical quantity, such as an object's position or temperature, changes over time. In fact, several sensors are necessary to complete the recommended activities, including an ultrasonic motion sensor, a force sensor, a temperature sensor, a pressure sensor, a voltage sensor, a magnetic field sensor, a rotary motion sensor, and a radiation sensor. Ideally, a computer will be available for every two students in a class. However, if fewer computers are available, students can work in larger groups or participate in interactive demonstrations provided by the instructor.

As of the writing of this edition, there are several companies that produce measurement sensors and data-acquisition systems. We briefly mention two of the more popular providers: *Vernier Science Education* (www.vernier.com) and *PASCO Scientific* (www.pasco.com). Each system typically involves software running on the computer, a wide variety of associated sensors, and a method for communicating between the sensor and computer. This communication can occur through an electronic interface device connecting the computer to the sensor, or it can be done wirelessly from the sensor to the computer with no intermediate device. Detailed information regarding the setup for each type of system and the use of the associated software are included in the system manuals and software help files.

In our classrooms, we generally use wired sensors and software from *Vernier Science Education*. As part of the software distribution, *Vernier* provides activity-specific files and templates for Workshop Physics. There are a few times in the activities when we refer to these files or to some aspect of the software, but we do our best to keep software-specific language to a minimum because we realize many instructors will be using different systems. The template files generally only pre-format the data collection and graphs, so they are not technically required. Also, because computer systems and hardware will continue to change in the future, instructors may need to provide a few basic instructions to get students going. Our experience is that after a couple tips to get started, students are very adept at figuring out how to use the data collection systems.

OTHER SOFTWARE

Analyzing motion in two dimensions can be greatly simplified by using video analysis software, and we make use of such software several times in the Activity Guide. Using such a program, position-time data can be quickly collected simply by clicking on the position of an object in each frame of a video. Most video analysis programs will also produce velocity and acceleration graphs and contain built in curve fitting (and mathematical modeling) routines as well. Both *Vernier Science Education* and *PASCO Scientific* offer video analysis tools; alternatively, there are several video analysis programs that are available as freeware.

We also note that two of our colleagues at Dickinson College, Lars English and Windsor Morgan, have developed a suite of simulations using the software package *VPython*. The simulations are designed to supplement materials for the introductory course and nicely complement the Workshop Physics Activity Guides. The materials are available as a free e-book entitled *VPython for Introductory Mechanics* by W. A. Morgan and L. Q. English.

OTHER TRADITIONAL PHYSICS EQUIPMENT

Because the investment in computer tools and sensors is substantial, we have tried to use standard physics equipment and inexpensive items that can be acquired locally for most of the activities. We assume that instructors will have access to basic equipment, including rods, clamps, metersticks, masses, stopwatches, scales, containers, rubber stoppers, etc. However, it should be noted that we use low-friction carts and tracks extensively when studying mechanics.

ASSIGNMENTS AND TEXTBOOK READINGS

After completing a set of activities, it is helpful for students to reinforce what they have learned by doing text-book readings and homework assignments. Although a traditional introductory physics textbook is not necessary when using the Activity Guide, we believe that such a resource is beneficial to students, particularly when seeking additional detail and clarification. In addition, such a resource provides a good source of homework problems for students. It is important that some of the homework problems focus on the types of activities and skills that students are learning in class. Thus, in addition to "traditional" textbook problems, instructors should assign problems that focus on a student's conceptual understanding of the material or their ability to analyze experimental data. Many textbooks and online homework systems now include these types of questions.

UPDATES TO THE THIRD EDITION

This third edition of the Activity Guide represents a significant revision. While the overall structure of the Activity Guide remains largely intact, we have tried to link the Units together in a more seamless and coherent manner. As part of this process, we now emphasize vectors from the very beginning and use vector notation throughout. Becoming comfortable with vectors and vector notation is a significant stumbling block for students, and we believe that repetition is the best way to overcome this barrier. We have also attempted to place a greater emphasis on some of the more fundamental aspects of physics, particularly Newton's second law, the Momentum Principle, and the Work-Energy Principle (and to a lesser extent the Rotational Momentum Principle). By emphasizing these fundamentals, especially surrounding concepts of energy, we hope that students will gain a more consistent and methodical understanding of, for example, how mechanics and thermodynamics are intimately connected.

Lastly, we have added new *Problem-Solving* sections throughout the Activity Guide that are typically located at the end of a Unit. The activities are designed to give students practice with more challenging problems that they can tackle in a group setting (ideally when an instructor is available to provide support). We believe such practice and support will ultimately result in more skilled and confident students. In addition, the *Problem Solving* sections can be collected and graded to provide additional, regular feedback to students.

ACKNOWLEDGEMENTS

ACKNOWLEDGEMENTS TO THE THIRD EDITION

The new co-authors (DPJ and BJP) are deeply indebted to Priscilla Laws for her longstanding work in physics education and the development of the Workshop Physics Activity Guides. Along with her numerous colleagues and collaborators, she helped pioneer the workshop approach at a time when it ran counter to the prevailing wisdom on how to teach physics. Her perseverance in developing not only the curriculum, but also many of the necessary computer-based tools, is what made Workshop Physics a success. We are honored to contribute to the continued development of the Activity Guide.

We are grateful to our colleagues and visitors in the Department of Physics and Astronomy at Dickinson College for their conversations and input, including Robert Boyle, Krsna Dev, Lars English, Catrina Hamilton-Drager, Windsor Morgan, and Hans Pfister. Jonathan Barrick helped create the experimental set-ups for many of the new activities in the Guide. The revised materials were undoubtedly improved by the students and teaching assistants in our introductory physics classes; their patience and willingness to contribute during the development phase is very much appreciated.

We also wish to thank the following individuals whose insights and feedback helped to improve the revised edition: David Baker (Austin College), Randy Booker (University of North Carolina at Asheville), Christopher Cline and Julia Kamenetzky (Westminster College), Danielle McDermott (Los Alamos National Laboratory), and Jeff Morgan (University of Northern Iowa).

Dickinson College continues to be supportive of the Workshop Physics program. It is satisfying to work at an institution that values pedagogical development and the student learning experience. We also thank Wiley for their long-standing work on the Workshop Physics Activity Guide and related materials. In particular, Jennifer Yee, Judy Howarth, and Samantha Hart have been extremely helpful (and patient) in shepherding the third edition through the publication process and kept us on track during the inevitable hitches and delays.

David Jackson and Brett Pearson
Department of Physics and Astronomy
Dickinson College
Carlisle, PA
October 2022

ORIGINAL ACKNOWLEDGEMENTS

All of us who were involved with this project owe a debt of gratitude to the Physical Science Study Committee for its pioneering work in the revitalization of introductory physics courses. Two individuals whose approach to physics teaching became popular in the 1960s deserve special mention for their insights into student learning difficulties—Robert Karplus of UC Berkeley and Eric Rogers of Princeton University. In addition, the work of Arnold Arons and Lillian McDermott of the University of Washington have provided inspiration for this work.

During the past eight years many people have contributed to the development of the Workshop Physics Project and this Activity Guide. First and foremost are the group of contributing authors: Robert Boyle (Units 16–18), Patrick Cooney (Unit 15), Kenneth Laws (Unit 25), John Luetzelschwab (Units 6–13 and 22–24), David Sokoloff (Units 3–7, 14, 16, 17, and 22–24), and Ronald Thornton (Units 3–7, 14, 16, and 17).

The following colleagues who have taught sections of the Workshop Physics courses at Dickinson College have contributed their insights based on the wisdom of experience. They include: Robert Boyle, Kerry Browne, David Jackson, Lars English, John Luetzelschwab, Windsor Morgan, Hans Pfister, Guy Vandegrift, and Neil Wolf. Hans Pfister deserves special mention for the design of kinesthetic apparatus for the Workshop Physics courses. In addition, several sabbatical visitors have helped in the development of activities including Mary Brown from

Dothan College, Desmond Penny from Southern Utah State College, and V. S. Rao from Memorial University in St. John's, Newfoundland.

The activities could not have been tested and refined without the work of several student generations of equipment managers and summer interns who helped during the early years of testing. They are Jennifer Atkins, Christopher Boswell, Joshua Clapper, Catherine Crosby, Ryan Davis, David Diduk, Christopher Eckert, Amy Filbin, Jake Hopkins, Michelle Lang, Mark Luetzelschwab, Despina Papazisis, Alison Sherwin, and Jeremiah Williams. I am also grateful to the 70 or so student assistants and graders and the approximately 700 students who have survived the Workshop courses as we tested and retested various activities.

Several Dickinson physics majors and project associates have developed software or software tools that have been used in the program including Grant Braught, David Egolf, Mike King, Sean LaShell, Mark Luetzelschwab, Brock Miller, and Phillip Williams. Several individuals in the Tufts University Center for Science and Mathematics Teaching who have rewritten early versions of computer-based laboratory software have redesigned portions of their software to meet our needs including Stephen Beardslee, Lars Travers, and Ronald Thornton.

The insights of colleagues from other departments and institutions have tested workshop activities or developed pedagogical approaches that have been helpful in the refinement of this Activity Guide. These colleagues are Nancy Baxter-Hastings (Dickinson College Department of Mathematics), Gerald Hart and Roger Sipson (Moorhead State University), Robert Morse (St. Alban's School), E. F. Redish and Jeffrey Saul (University of Maryland), Mark Schneider (Grinnell College), Robert Teese (Muskingham College), Maxine Willis (Gettysburg High School), William Welch (Carroll College), and Jack Wilson (Rensselaer Polytechnic Institute). Early adopters who have made contributions include Mary Fehrs and Juliet Brosing at Pacific University, Bruce Callen at Drury College, Ted Hodapp at Hamline College, Jim Holliday at John Brown University, Bill Warren at Lord Fairfax Community College, Bill Wehrbein at Nebraska Wesleyan, and Maxine Willis at Gettysburg High School.

Several administrators at Dickinson College have arranged for financial support for purchasing equipment, remodeling our classroom and equipment storage areas, and providing facilities for project staff. These individuals include President A. Lee Fritschler, Deans George Allan and Margaret Garrett, the treasurer, Michael Britton, and grants officer Christina Van Buskirk.

Individuals from the commercial sector have helped with the design, production, and distribution of hardware, software, and apparatus needed for the activities in this guide. They include: David and Christine Vernier of Vernier Software and Technology, Paul Stokstad and David Griffith of PASCO Scientific, Rudolph Graf of Science Source, and Ron and Wendy Budworth of Transpacific Computer Company.

Workshop Physics Project support staff who have helped with the production of this Activity Guide include Susan Greenbaum, Gail Oliver, Susan Rogers, Pam Rosborough, Sara Settlemyer, Virginia Trumbauer, Maurinda Wingard. Kim Banister, Erston Barnhart, Kevin Laws, Virginia Jackson, and Noel Pixley have helped with the artwork. Wiley editors Clifford Mills and Stuart Johnson with the help of Katharine Rubin and Geraldine Osnato have coordinated the Activity Guide production effort.

Major support for this work was provided by the Fund for Improvement of Postsecondary Education (Grants #G008642146 and #P116B90692-90) and the National Science Foundation (Grants #USE-9150589, #USE-9153725, #DUE-9451287, and #DUE-9455561). Project Officers who have provided administrative and moral support for this project are Rusty Garth, Brian Lekander, and Dora Marcus from FIPSE, and J. D. Garcia, Ruth Howes, Kenneth Krane, and Duncan McBride from NSF.

Priscilla Laws
Department of Physics and Astronomy
Dickinson College
Carlisle, PA
January 2004
On behalf of contributing authors Robert Boyle, Patrick Cooney,
Kenneth Laws, John Luetzelschwab, David Sokoloff, and Ronald Thornton

UNIT 19: ELECTRIC FORCES AND FIELDS

Ken Kaminesky/Stockbyte/Getty Images

This person's hair sticks out when they place their hand on a Van de Graaff generator. What is the generator doing that makes their hair stand on end? After you study the fundamental nature of electrical forces in this unit, you should be able to answer this question.

UNIT 19: ELECTRIC FORCES AND FIELDS

OBJECTIVES

1. To explore the basic properties of electric charge.

2. To understand how Coulomb's law describes the forces between charged objects.

3. To become familiar with the concept of an electric field.

4. To learn how to calculate the electric field associated with a distribution of charges.

19.1 OVERVIEW

You've probably noticed that it's possible for one object to "stick" to another despite there being no tape or glue between them. For example, rubbing a balloon on your hair may allow it to stick to a wall or to pick up small pieces of paper. Or perhaps your clothes have suffered from "static cling," and you ended up with a sock sticking to your shirt. We begin our study of electrical phenomena by exploring the nature of forces between objects due to a fundamental property known as *electric charge* (or simply *charge*). When the objects are not moving (or moving very slowly), this interaction is known as the *electrostatic force*.

We start by investigating the circumstances under which electrostatic forces are either attractive or repulsive. We then consider how the forces between charged objects depend on the (net) charge of the objects and the distance between them. This will lead to a formulation of *Coulomb's law*, a mathematical relationship that expresses the electrostatic force between two objects.

Finally, we will define a quantity called the *electric field*. The electric field is analogous to the gravitational field in that it provides a "picture" of the force vectors. Given an arbitrary distribution of charges, one can determine the electric field at *any* point in space. Such a concept provides a powerful tool for understanding the nature of electrostatic forces.

ELECTROSTATIC FORCES

19.2   EXPLORING THE NATURE OF ELECTRICAL INTERACTIONS

We begin by investigating some basic properties of electrical interactions. Each group should have the following equipment available:

- Several pieces of Scotch tape
- 2 rod stands, metal rods, and 90° clamps
- 2 low-mass insulating objects hanging from non-conducting thread (e.g., small Styrofoam balls or table-tennis balls)
- 2 low-mass conducting objects hanging from non-conducting thread (e.g., small pith balls, crumpled aluminum foil, or table-tennis balls coated with conducting paint)
- 1 hard plastic rod (e.g., PVC tubing)
- 1 piece of fur
- 1 glass rod
- 1 polyester cloth

Two Types of Charge

Even if you haven't studied electrostatics formally, you may have heard that there are two types of charge and that charged objects can exert forces on each other. However, as you explore these initial activities, try to approach the experiments with a "clean slate." In other words, you should base your conclusions on experimental observations as opposed to your prior knowledge (though we realize this is not always easy to do).

19.2.1. Activity: Interactions with Scotch Tape

a. Take *two* pieces of Scotch tape, fold over a small portion of each so that you have a non-sticky portion to grab, and then stick both pieces flat on the table. Quickly pull both pieces up off the table (being careful to keep them separated), and then slowly bring them close to each other and observe what happens. It works well if you hold the pieces so they dangle below your hands, bringing your hands close together. (Try not to let the pieces touch each other.) What do you see? Do they interact, and if so, in what way? Does the distance between the pieces appear to affect the interaction? Briefly explain.

b. Now get *four* new pieces of tape (probably best for two people to each get two). Take two of the pieces and stick them to the table, just like in part (a). Write a B on each piece to label them as "bottom." Then take the two remaining pieces and stick them directly on top of each of the B pieces; label these strips T for "top." Note that these pieces should be stuck directly to the backs of the B pieces and should not be touching the table.

One person should grab *one* of the B pieces and pull up one B and T pair off the table as a single piece. A second person should do the same with the other pair. You should each be holding two pieces of tape that are stuck together. Now, each person should *quickly* pull their two pieces of tape apart, being careful to keep them from touching anything (or each other once they are separated). The following questions direct you to bring the pieces close to each other *pair-wise* and observe what happens with each pair, trying not to let them touch.

1. Describe the interaction between two *top* pieces ("T") when they are brought close to one another. Does the interaction appear to depend on how far apart the two pieces are?

2. Similarly, describe the interaction between two *bottom* ("B") pieces.

3. Finally, describe the interaction between one *top* and one *bottom* piece.

c. Describe how your observations can be explained assuming there are two types of charge (say, top and bottom).[1] Don't simply state this as something you already know, but instead explain how your three observations from part (b) show this to be true.

d. As we will see, there are two types of charge that are labeled positive and negative. Based only on your observations, is there any way to tell

---

[1] The observations you just made cannot be explained using only one type of charge; it requires at least two. However, we have only shown that there are *at least* two types of charge. It is conceivable that some other physical process could produce a third type of charge (although given that the interactions appear to attract or repel, it is not clear what type of interaction this third charge would have). As far as scientists know, there are only two types of electrical charge.

which piece of tape (top or bottom) is positively charged and which is negatively charged? Briefly explain.

What you just observed involves a series of electrical interactions due to a property of matter we call *electric charge*.[2] There are two types of charge, arbitrarily labeled *positive* and *negative*, and as you just observed, one can produce charged objects simply by peeling a piece of Scotch tape off a table. You should have found that the two bottom pieces *repel* each other, as do the two top pieces. In addition, the interactions get stronger as the pieces are brought closer together. On the other hand, a bottom piece and a top piece are *attracted* to each other.

While we can conclude that there must be (at least) two types of charge, we cannot determine which piece of tape has which type of charge (positive or negative). Benjamin Franklin *arbitrarily* assigned the term "negative" to the type of charge that resides on a hard plastic rod (or, in his day, a rubber rod) when rubbed with a piece of fur. One can produce the opposite kind of charge ("positive") by rubbing a glass rod with a piece of silk or polyester. Note this choice of positive and negative is purely arbitrary; the term "negative" could just as well have been assigned to the charge on the glass rod. It wasn't until scientists developed a *microscopic* understanding of matter that one could determine the underlying cause of these different types of charge.

Conductors, Insulators, and Polarization

The matter we encounter on a daily basis is made of atoms that contain both positive charges (associated with the protons in the nucleus) and negative charges (associated with the electrons). Most of the time, electrons in an atom surround an equal number of protons, so the net charge is zero. In this case the atom is said to be *electrically neutral* (also referred to as *uncharged*).

In some types of solid materials, known as *insulators*, the electrons are tightly bound to the protons in the nucleus and do not easily move away from their host atoms. In other solids, known as *conductors*, some electrons are weakly bound and free to move between atoms due to the influence of other charges. These microscopic effects lead to differences in macroscopic behaviors between the two types of materials.

In the next set of activities, we will explore these types of interactions in an effort to understand the underlying principles. We will create charged objects by rubbing fur against a plastic rod (or rubbing polyester against a glass rod).

Benjamin Franklin

---

[2] A property of matter is not the same thing as the matter itself. For instance, a full balloon has several properties at once—it can be made of rubber or plastic, have the color yellow or blue, have a certain surface area and so on. Thus, we don't think of charge as a substance but rather as a property that certain substances can have at times. That being said, it's common to causally refer to excess charge as if it were a substance, so don't be misled by this practice (that we will likely indulge in over the next few units).

Before rubbing a plastic rod with fur, the plastic is (essentially) neutral. But after rubbing it with fur, some extra electrons from the fur remain on the rod, and the rod ends up with an overall negative charge.[3]

The nature of electrical interactions is not obvious without careful experimentation. But be warned: charge is easily transferred, and unintentionally touching an object will likely remove the charge (as can humid air). Unfortunately, this means there's a chance that an experiment won't always work perfectly on the first try. When this happens, just try the experiment again, being as careful as you can.

19.2.2. Activity: Conducting Objects

a.   Start by hanging a low-mass conductor (e.g., a small pith ball) from the stand. Touch the object with your hand to make sure it is not charged. Next, vigorously rub a plastic rod with a piece of fur for a few seconds and then *slowly* bring the rod close to the object while watching carefully to see what happens. Explain what you observe below. **Note:** The process happens quickly, so you may need to try it a few times to understand what's going on (make sure you discharge the object by touching it with your hand before each attempt).

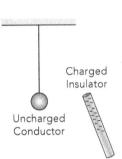

Charged Insulator

Uncharged Conductor

   1. *Before* the two objects touch, what do you observe when the negatively-charged rod is brought near the uncharged conductor (attraction, repulsion, neither)? If there is an interaction, is it strong or weak?

   2. Let the rod touch the conductor. Then once again bring the negatively-charged rod near the conductor. *After* the two objects have touched, what do you observe when they are brought close together (attraction, repulsion, neither)? If there is an interaction, is it strong or weak?

b.   You may find this result surprising. Let's begin by thinking about observation (1). You should have seen that *before* the objects touch, the *neutral* conductor is (strongly) *attracted* to the negatively-charged rod. Use your observations from Activity 19.2.1, along with the fact that charges can move easily on conductors, to explain why a neutral conducting ball is attracted to the rod before touching it. **Hint:** Draw a diagram showing

---

[3] Because those electrons come from the fur, the fur ends up with a deficit of electrons and therefore becomes positively charged.

the location of charges on the objects, remembering that the conductor is neutral overall.

c. Now consider observation (2). *After* the objects have touched, the conductor is (strongly) *repelled* from the negatively-charged rod. Use your observations from Activity 19.2.1 to explain why the conducting ball is repelled from the rod after touching it. Once again, it's helpful to draw a diagram showing the location of charges on the objects.

There are two major takeaways from this activity. First, a neutral (uncharged) conductor is strongly attracted to a charged object. Although the conductor is neutral overall, it is made up of a huge number of atoms (or molecules) that contain both positive and negative charges. In a conductor the microscopic charges are free to move around, and in this case the electrons in the conductor are repelled by the negatively-charged rod and move as far away as they can (to the far side of the object). The result is that the portion of the object *closest* to the charged rod is left with a net positive charge, and this positive charge is attracted to the negatively-charged rod.

It's important to note that the conductor remains electrically neutral overall (the net charge on the object remains zero), and both attractive and repulsive forces are present. It's the *distribution* of charge on the conductor that changes, and since the positive charges are closer to the (negatively-charged) rod than the negative charges, the magnitude of the attractive force is larger than that of the repulsive force. Make sure you fully understand this argument before moving on!

The second major takeaway is that a conductor is strongly repelled by a charged object after touching it. This repulsive force suggests that the conductor is no longer neutral, and that some of the negative charges from the rod were transferred to the conductor so that it now has a net negative charge. The two negatively-charged objects then have a repulsive force between them.[4] Thus, another feature of conductors is that they readily accept charge—it's easy to "charge a conductor."

---

[4] There is a subtle issue here. Charges are free to move on the conductor, and negative charges will be repelled by the negatively-charged rod and tend to move to the far side of the conductor. So, it's possible that the conductor might still be attracted to the negative rod even if it has a net negative charge. But this will only happen if the conductor has a *very small* net charge. It is almost always the case that a significant amount of charge is transferred to the conductor, so the overall force is typically repulsive.

19.2.3. Activity: Insulating Objects

a. Now hang a low-mass *insulator* (e.g., a small Styrofoam ball) from the stand. Touch the object with your hand to make sure it is not charged. Use a piece of fur to (negatively) charge a plastic rod and then *slowly* bring the rod close to the object while watching carefully to see what happens.

    1. *Before* the two objects touch, what do you observe when the negatively-charged rod is brought near the insulator (attraction, repulsion, neither)? If there is an interaction, is it strong or weak?

    2. Let the rod touch the insulator. Then once again bring the negatively-charged rod near the insulator. *After* the two objects have touched, what do you observe when they are brought close together (attraction, repulsion, neither)? If there is an interaction, is it strong or weak?

b. Consider observation (1). *Before* the objects touch, the neutral insulator is (weakly) *attracted* to the negatively-charged rod. Although we defined an insulator to be a material in which the electrons cannot readily move around (they remain bound to their host atom or molecule), on a *microscopic* scale the charge distribution in the atom or molecule might deform slightly. Use this idea to explain why the (neutral) insulating ball is attracted to the negatively-charged rod before touching it. Once again, a picture will be helpful!

c. Now consider observation (2). *After* the objects have touched, the insulator is still (weakly) *attracted* to the negatively-charged rod. Does the insulator appear to have picked up any charge when the objects touched? Explain what you think is happening.

You should have seen that the behavior of an insulator is different from a conductor in two main ways. First, although a neutral insulator is attracted to a charged object (just like a conductor), the effect is much weaker. Although the microscopic charges in an insulator are not free to leave their host atoms, the positive charges in the atoms will still be attracted to the (negatively) charged

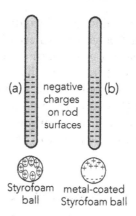

Fig. 19.1. Depiction of how a (negatively) charged object influences a neutral insulator and a neutral conductor. (a) The electrons in the insulating ball are repelled from those in the rod but stay with their host atoms, leading to a *microscopic* polarization. (b) Electrons in the conducting ball are repelled from those in the rod and are free to move as far away from the rod as possible, leading to a *macroscopic* polarization. (The plastic rod is an insulator, and so the electrons on it stay fixed.)

rod while the negative charges will be repelled. Thus, the atoms (or molecules) will have an asymmetric charge distribution, with the positive charges moving slightly closer to the rod and the negative charges moving slightly farther away.[5] This separation of charge is referred to as *polarization* and is depicted schematically in Fig. 19.1 (a).

Recall that in a conductor the negatively-charged electrons are free to move to the opposite side of the object. This means that ball has a net negative charge bunched on one side, leaving a net positive charge on the other. This process is quite similar to what happens at the microscopic scale of an insulator, but because it happens on a *macroscopic* scale, it is known as a *macroscopic polarization* and is shown in Fig. 19.1 (b). Note that the charges in a conductor end up being separated by much larger distances than in an insulator, and this leads to the (neutral) conductor being more strongly attracted to a charged object compared to the (neutral) insulator.

The second way the insulator behaves differently from the conductor is that it remains attracted to the charged rod, even after touching it. When the charged rod touches the conducting ball, the negative charges are readily accepted by the ball because the negative charges on the conductor have already moved to the other side. Charges are easily transferred to a conductor, leaving the ball negatively charged as well. Because charges are not free to move around in an insulator, few, if any, negative charges are transferred to the object. The insulator remains essentially neutral overall, so it is still attracted to the charged object even after touching it.

It is important to note that an insulator *can* become charged. In fact, the plastic rod used in the previous experiment is an insulator, and it becomes charged after rubbing it with fur. The point is that a conductor is *easy* to charge—you simply need to touch it with a charged object, while an insulator takes a fair amount of effort to charge—you need to repeatedly rub it for it to accept charge (and that charge will then remain fixed in place).

---

[5] Although this charge separation is *tiny*, there are a very large number of atoms (or molecules) making up the object!

Finally, it's worth mentioning that Activities 19.2.2 and 19.2.3 can be repeated using a glass rod and polyester (instead of the plastic rod and fur). In this case the rod will be charged positively instead of negatively, but all the observations remain the same. It turns out that a glass rod tends to be more difficult to charge than a plastic one, so the interaction is typically stronger when using plastic and fur. (If you want to verify that the glass rod gets the opposite charge as the plastic rod, try charging the small conducting object using a charged plastic rod and then see what happens when you bring up a charged glass rod.)

19.3 FORCES BETWEEN CHARGES—COULOMB'S LAW

Coulomb's law is a mathematical description of the fundamental nature of the electric force between charged objects that are either spherical in shape or small compared to the distance between them (so that they act like point particles). Unfortunately, determining Coulomb's law experimentally is not easy, mainly because it takes extreme care to maintain a fixed amount of charge on an object. Therefore, our approach will be as follows. We begin by *predicting* how the charge on two objects and their separation affect the mutual force between them. We will then state Coulomb's law mathematically and work through a few examples to become familiar with how to use it. In a later activity, we will quantitatively verify Coulomb's law using video analysis.

To quantitatively investigate the forces acting between charged objects, Coulomb devised a clever trick so that he would not have to know the actual amount of charge on the objects. He transferred an unknown amount of charge q to a conductor, and then touched the newly charged conductor to an identical, uncharged one. In such a situation, the conducting objects will end up with the *same* amount of excess charge, in this case each ending up with a charge of $q/2$. After observing the interactions with this amount of charge, Coulomb discharged one of the conductors and then touched the conductors together again, giving each a charge of $q/4$, and observed their interactions. He repeated the procedure with a charge of $q/8$, $q/16$, and so on. Working systematically in this manner, Coulomb was able to determine the precise form of the force law acting between charged objects.

Consider a pair of conductors (labeled A and B) hanging from strings, each with an initial excess charge of $q_A = q_B = q/2$, where q is some unknown amount of charge (as shown in Fig. 19.2). Because there is a repulsive force between the conductors, they will hang at a particular angle with respect to the vertical when placed a certain distance apart. In the following activity you will predict what happens to this angle as the experimental parameters are varied. In each case, give a reason for your prediction.

Fig. 19.2. Original position of the two charged hanging conductors.

19.3.1. Activity: Predicting Dependence of Force on Charge and Distance

a. Assume the two stands in Fig. 19.2 are moved *closer together*. How do you think the force between the objects will change, and how will this alter the angle of the strings with the vertical (as compared to the original angle in Fig. 19.2)? Explain briefly.

b. Now assume the stands are moved back to their *original* position, but the amount of charge on each conductor is *decreased* to $q_A = q_B = q/4$. How do you think the force between the objects will change, and how will this alter the angle of the strings with the vertical (as compared to the original angle in Fig. 19.2)? Explain briefly.

c. Based on your predictions above, how do you think the force law between charged objects should depend on the charges q_A and q_B, and on the distance r between the charges? (We understand that you are making a bit of a guess here, but you should at least be able to determine whether these quantities should appear in the numerator or the denominator.)

d. In what direction do you think the electrostatic force acts on each object? Make a rough sketch similar to Fig. 19.2, and then draw in the electrostatic force vectors acting on each object. **Hint**: Remember Newton's third law!

Coulomb's law asserts that the magnitude of the force between two electrically-charged (spherical) objects is *directly proportional to the product of the amount of excess charge on each object and inversely proportional to the square of the distance* between their centers. The direction of the force is along a line between the two objects and is attractive if the particles have opposite signs and repulsive if they have like signs. All of this is expressed in Eq. (19.1) below, which gives the electrostatic force exerted *by* object A *on* object B (which we abbreviate as $A \rightarrow B$). Equation (19.1) is known as *Coulomb's law* (note that the force direction is given by the unit vector $\hat{r}_{A \rightarrow B}$.

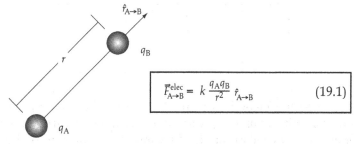

$$\vec{F}^{\text{elec}}_{A \rightarrow B} = k \frac{q_A q_B}{r^2} \hat{r}_{A \rightarrow B} \qquad (19.1)$$

Fig. 19.3. Diagram showing the direction of the unit vector $\hat{r}_{A \rightarrow B}$ used in the Coulomb force equation that describes the force on charge q_B due to charge q_A.

In Eq. (19.1) k is known as Coulomb's constant (given by 9.0×10^9 N $\cdot$ m^2/C^2); q_A and q_B represent the charges on objects A and B, respectively (measured in *coulombs*, the SI unit of charge); r is the distance between the centers of the two objects; and the symbol $\hat{r}_{A \to B}$ (pronounced r-hat) represents a *unit vector* directed from object A toward object B. Because the charges q_A and q_B can be positive or negative, Eq. (19.1) may end up with a negative sign out front, causing the vector to point in the opposite direction.

For many students, the most difficult part of Eq. (19.1) is understanding the meaning of the unit vector $\hat{r}_{A \to B}$. Because of this, some students prefer to remember that the magnitude of the force is given by $k|q_A||q_B|/r^2$, and then use the fact that the force is either repulsive (like charges) or attractive (unlike charges) to find the direction. The following activity is designed to help you become more familiar with Coulomb's law.

19.3.2. Activity: Understanding Coulomb's Law

a. In the diagram below, draw the unit vector $\hat{r}_{A \to B}$ (don't worry too much about the length of the vector, the direction is the important part). Its tail should be located on q_B, and it should point along the direction from object A toward object B. Also, draw the unit vector $\hat{r}_{B \to A}$ with the tail on object A (this is the unit vector that would appear in the equation for the force of object B on object A: $\vec{F}_{B \to A}$).

b. What do you notice about the unit vectors $\hat{r}_{A \to B}$ and $\hat{r}_{B \to A}$? How are they related to each other? Can you write down a mathematical expression for how they are related?

c. Coulomb's law contains the product $q_A q_B$. In the table below, indicate the sign of the *product* $q_A q_B$ for each combination of positive and/or negative charges.

Sign of q_A	Sign of q_B	Sign of $q_A q_B$
+	+	
+	−	
−	+	
−	−	

d. According to Coulomb's law, the force vector involves a product $q_A q_B$ and the unit vector $\hat{r}_{A \to B}$. Use your responses to parts (a) and (c) to draw an arrow indicating the direction of the force exerted on object B by object A if the charges are both positive or both negative. As usual, place the tail of the vector on object B. (Does your arrow agree with the statement "like charges repel"?)

e. Similarly, use your previous responses to draw an arrow indicating the direction of the force exerted on object B by object A if the charges have opposite signs (that is, one is positive and one is negative). Again, place the tail of the vector on object B. (Does your arrow agree with the statement "unlike charges attract"?)

f. Write down an expression similar to Eq. (19.1) for the force of object B on object A, and then use your answer to part (b) to show that the two forces $\vec{F}_{A \rightarrow B}$ and $\vec{F}_{B \rightarrow A}$ obey Newton's third law.

Mathematically, the direction of the force on object B comes from the product $q_A q_B \hat{r}_{A \rightarrow B}$ in Eq. (19.1) (with a similar expression for the force on object A). As mentioned, many people find it easier to remember that the force always lies along the line between the two charges and is either repulsive or attractive, depending on whether the charges are the same or opposite. You can use whichever method you prefer, but it's important to remember that the electrostatic force, like all forces, is a vector, and so it needs to be expressed using proper vector notation (either in components or as a magnitude and direction).

The following activity will give you some additional practice with Coulomb's law and how to determine the direction of the electrostatic force. **Note**: A coulomb is a *very large* amount of charge, so it is common for the amount of charge on objects to be given in the nanocoulomb (nC) or microcoulomb (μC) range.

19.3.3. Activity: Using Coulomb's Law

a. Consider two point-like, charged objects that lie along the x-axis. Object A has a charge $q_A = 2.0 \times 10^{-9}$ C and is located at $x = 3.0$ cm, while object B has a charge $q_B = -3.0 \times 10^{-9}$ C and is located at $x = 5.0$ cm. Draw a diagram showing the two charges and then determine the force on the negative charge (q_B) due to the presence of the positive charge (q_A).

Express the force two different ways: (1) using components and unit vectors, and (2) as a magnitude and angle with respect to the positive x-axis.

b. Now suppose q_B is moved to the point (5.0 cm, 6.0 cm). Once again, draw a diagram showing the two charges and then determine the force on q_B due to the presence of q_A. Express the force two different ways: (1) using components and unit vectors, and (2) as a magnitude and angle with respect to the positive x-axis.

QUANTITATIVE ASPECTS OF COULOMB'S LAW

19.4 VERIFYING COULOMB'S LAW

In this section we seek to quantitatively verify Coulomb's law using the following method. A small, conducting sphere is placed on the end of an insulating rod and charged negatively using a plastic rod that has been rubbed with fur. This sphere, having charge q_A, is then used as a prod to push on another sphere of charge q_B (and mass m) that is suspended by a thread. As q_A is moved closer, the angle of the thread with the vertical will increase (see Fig. 19.4). Using a video of this situation allows us to determine the angle of the suspended object, as well as the distance r between the prod and the suspended object. This information will allow us to determine experimentally how the force depends on the distance between the two objects.

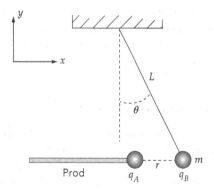

Fig. 19.4. Diagram showing the information needed to determine the forces on object B.

19.4.1. Activity: Forces on a Suspended Charged Object—Theory

a. Begin by drawing a free-body diagram for object B in the space below using Fig. 19.4 as a guide. For now, just write the magnitude of the electrostatic force as F^{elec} instead of using Coulomb's law. **Hint**: There are three forces acting.

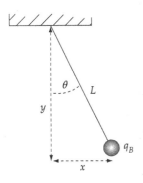

Fig. 19.5.

b. Assuming everything is at rest in Fig. 19.4, use Newton's second law in both the x and y directions to solve for the magnitude of the electrostatic force F^{elec} in terms of the mass m, the angle θ, and the gravitation field strength g.

c. Next, use the geometry of Fig. 19.5 to express F^{elec} solely in terms of m, g, x, and L. **Hint**: Start by finding the distance y as a function of x and L, and use this to eliminate $\tan\theta$ in your result from part (b).

You should have found that magnitude of the electrostatic force in this situation can be written as $F^{\text{elec}} = mgx/\sqrt{L^2 - x^2}$. This expression allows us to use the experimentally-measured values of m, x, and L to determine the magnitude of the electrostatic force F^{elec}. After we analyze the experiment, we will use this result to connect back to Coulomb's law to determine how this force depends on the distance between the two objects.

To proceed, we need a video of the actual experiment. If you are making your own video, follow the instructions below. Otherwise, your instructor may ask you to analyze a pre-recorded experiment, in which case you can skip the following activity.

19.4.2. (OPTIONAL) Activity: Recording the Coulomb's Law Experiment

To perform this experiment, you will need the following equipment:

- 1 conducting, low-mass ball threaded on a string (e.g., metal-coated table-tennis ball)
- 2 non-conducting threads, 1–2 meters long
- 1 rod (or pair of ceiling hooks), for suspending the ball
- 1 prod (metal conducting ball with an insulating handle)

- 1 hard plastic rod (e.g., PVC tubing)
- 1 piece of fur
- 1 electronic balance (for finding the mass of suspended ball)
- 1 ruler or meter stick

Setting up the Force-Law Experiment

The purpose of this experiment is to verify that the magnitude of the electrostatic force between two small objects varies as $1/r^2$, where r is the distance between the centers of the balls.

1. Suspend the conducting ball from a long, non-conducting thread (for lateral stability, two pieces of thread should be used, as shown in Fig. 19.6). Record the vertical distance from the point of suspension to the center of the hanging sphere. Place a meter stick horizontally under the hanging ball to set the scale.
2. Use fur to charge a plastic rod and use it to charge the prod.
3. Carefully touch the ball with the prod so that the two objects have the same amount of charge.
4. Slowly and steadily bring the charged prod closer and closer to the hanging ball, while *keeping the line between the ball and the prod horizontal at all times*. This step might take some practice.
5. Once you get good at step 4, perform the experiment while recording a video. Start the movie with the prod far enough from the ball so that there is no noticeable interaction between the two, and be sure to move slowly.

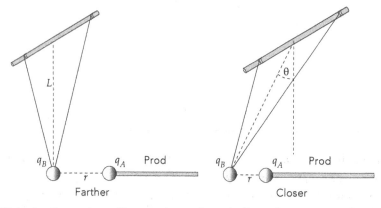

Fig. 19.6. An illustration of how a charged prod affects the position of a charged conducting ball hanging from two threads (which adds lateral stability to the hanging ball).

In the next activity, we analyze the recorded video of the experiment. Our goal is to see whether the experimental data agree with the form of the electrostatic force in Coulomb's law.

19.4.3. Activity: Verifying Coulomb's Law Experimentally

a. Begin by using the video analysis software to scale the video and set the origin (think about which point might make a convenient origin). Next, record the positions of both balls (the prod and the hanging ball) in every frame. Then use the software to determine the following two quantities in each frame of the video: (i) the separation distance r between the two charged objects, and (ii) the horizontal distance x from the suspended ball to the vertical line where it started. You should also measure the length L of the pendulum if not provided.

b. Based on the equation you derived in Activity 19.4.1 part (c), use the software to calculate the electrostatic force magnitude at each frame in the video.

c. Finally, plot the electrostatic force as a function of the separation distance r. Try fitting your data using both A/r and A/r^2, where A is a constant. You should find that one of these works much better than the other. Write down your best fit equation below and comment on whether the dependence on r agrees with Coulomb's law. If the agreement is not great, describe what you think are the most plausible sources of uncertainty in the experiment. Print out (or sketch) your graph with fit below.

d. Because the prod touched the ball before the experiment, the two balls should each have the same amount of charge. In other words, $q_A = q_B \equiv q$, where q is the unknown amount of charge on the prod and suspended ball. This means we can write the magnitude of the electric force as

$$\left|F^{elec}\right| = \left|\frac{kq_Aq_B}{r^2}\right| \rightarrow \frac{kq^2}{r^2}$$

Choose a specific value of the distance r and use this expression to calculate the actual amount of charge q on the balls in your experiment. (It will likely be much smaller than one Coulomb!)

19.5 ELECTROSTATIC DEMONSTRATIONS

In addition to exploring the nature of the relatively small collections of electric charge that result from rubbing objects together, we can also look at demonstrations that involve a much larger amount of electric charge. For the demonstrations in this section, you will need:

- 1 Van de Graaff generator, including a grounding sphere
- Addition items such as a wig, metal pie plates, an all-metal thumbtack with tape, and/or a small conducting sphere that can be suspended next to the Van de Graaff generator.

19.5.1. Activity: Van de Graaff Demonstrations

The Van de Graaff Generator

Ben Franklin and others recognized that electric charge can be "produced" by doing mechanical work (such as rubbing a plastic rod with fur). The Van de Graaff (VdG) generator is a device that uses a moving belt to deposit electric charge on a large, hollow metal sphere. Several interesting demonstrations can be performed with this device, including:

- Hair raising—If you place your hand on the VdG and turn it on, charges will cover your body in addition to the metal sphere. These charges repel each other and end up covering your entire body, including your hair. When straight hair is covered with charge, it will begin to stand straight up (and out) as the individual strands are repelled from each other. **Caution**: You need to be insulated from ground when performing this demonstration. Be careful to discharge yourself properly or you will receive an uncomfortable shock. This demonstration is safer to do (and more impressive) by placing a wig directly on the VdG sphere.
- Flying pie plates—If we place a stack of metal pie plates on top of the VdG sphere and turn it on, the plates will all become charged and repel each other. You should see the pie plates start to slowly float off the VdG as the electrostatic force becomes large enough to overcome the gravitational force.
- Sudden electrical discharge ("lightning")—Because like charges repel, the charges being supplied to the sphere of the VdG generator repel each other and spread out all over the sphere. If we bring a grounded (neutral) conductor near the VdG sphere, the charges will be attracted to it. If the attraction is strong enough, the charges will "jump" across the gap, creating a spark in the air. (The charges don't actually jump, of course. The large amount of charge actually ionizes the air, literally ripping electrons from the air molecules and creating a plasma. The resulting charges in the air allow the excess charges on the VdG to "flow" through the plasma along the path of the spark. The visual spark that is seen is due to the electrons re-combining with the ions, emitting light in the process.)
- Lightning rods—The main purpose of a lightning rod is to protect a building from a lightning strike. The lightning rod is attached to the highest point on the structure and provides a safe, alternate path to ground (as opposed to through the building) in the event of a lightning strike on the building. Interestingly, if we repeat the lightning experiment above but first tape a thumbtack to the metal sphere to mimic a lightning rod, then the dramatic sparks are actually eliminated. The reason for this is because as the charges spread out on an object, they get pushed toward the edges and sharp points. Thus, charges will build up on the point of the thumbtack much more quickly than on the sphere itself, which allows for charge to "leak" off the thumbtack, preventing the charge from building up to dangerous levels. Therefore, in addition to protecting a building in the event of a lightning strike, a lightning rod can help prevent a strike in the first place. (In an actual thunderstorm the

charge buildup can be so immense that a lightning rod cannot prevent all lightning strikes.)

- Bouncing ball—If a conducting ball hanging from a thread is placed between the VdG sphere and a grounded sphere, the conducting ball will bounce back and forth between the two spheres as it continually transfers charge between the two spheres.

Your instructor will perform one or more of the above demonstrations. In the space below, briefly explain the demonstration(s), what you saw, and why it occurred.

THE ELECTRIC FIELD

19.6   THE ELECTRIC FIELD

Up until now, most of the forces we have studied are the result of direct action or "contact" of one piece of matter with another. For example, we have pushed on a cart with our hands, tied a mass to a string, observed the table exert a normal force or a friction force, etc. The one obvious exception to this is the gravitational force, which can clearly act with no actual contact between two objects. From your observations of charged objects earlier in this unit, it should be clear that electrical forces can also act without direct physical contact between the objects. The "action at a distance" that characterizes electrical and gravitational forces is, in some ways, inconceivable to us. How can one object feel the presence of another object with only empty space in between?[6]

It turns out that what we normally think of as everyday "contact" forces (e.g., normal, tension, etc.) are more complicated when you look at them from the microscopic level. For example, what actually provides the contact force when you push your hand down on the table? All atoms and molecules are thought to contain electrical charges, and physicists believe that these "contact" forces are nothing more than electrical forces involving very small separations. In other words, it is the electrical charges contained in the molecules in your finger and the table top that end up repelling each other and preventing your finger from passing right through the surface. So, even though "contact" forces seem quite different from electrical forces acting over a (relatively large) distance, they are really one in the same.

Historically, many physicists were intrigued by forces that could act through a distance. Let's consider the attempts of Michael Faraday and others

---

[6] Modern physical theories, such as quantum electrodynamics, hypothesize that such forces are actually the result of particles being exchanged between the two objects.

to explain such action-at-a-distance forces during the nineteenth century. Understanding more about these attempts should help you develop some useful models to describe the forces between charged objects.

To describe action at a distance, Michael Faraday introduced the notion of an *electric field* emanating from a collection of charged objects and extending out into space. More formally, the electric field due to a known collection of charged objects is represented by an electric field vector at every point in space. This electric field will in turn affect any charges in the region. One typically thinks of putting a "test charge" (typically a very small, positive charge) at some point in space and looking at the force it experiences. The electric field is then defined as the force per unit test charge; in this way, the actual value of the test charge plays no role. The electric field due to a collection of source charges is therefore defined by

$$\vec{E}_s \equiv \frac{\vec{F}_{s \to t}}{q_t} \tag{19.2}$$

where q_t is the charge of the test particle and $\vec{F}_{s \to t}$ is the net force on the test charge due to all the source charges. Note that since $\vec{F}_{s \to t}$ will, in general, change as you move the test charge around, the electric field vector will also (in general) have a different magnitude and direction at each point in space.

Finally, assuming we know the electric field, we can write the force in terms of the field (where we have switched notation to a general charge q):

$$\vec{F}_{s \to q} = q\vec{E}_s \tag{19.3}$$

Because the electric field is independent of the test charge, this expression gives the force for *any* charge q. Equation (19.3) says that if you know the electric field due to a collection of source charges, you can very easily find the force on any charge simply by multiplying the charge by the field! Note the force on the test charge points in the same direction as the electric field $\vec{E}_s$ (or opposite if q is negative). Equation (19.3) is extremely useful when finding the force on a charge due to an existing electric field.

Electric Field of a Point Charge

To make things more concrete, let's begin by determining the electric field due to the simplest possible collection of sources—that of a single "point" charge. Consider a point particle that has a charge q_p and is fixed in space (we will begin by assuming the charge is positive, but it works either way). Now imagine placing a small, positive test charge q_t somewhere in the vicinity of the charge q_p. Figure 19.7 shows the idea, where the test charge is represented by the small, black dots that can be placed at various locations around the fixed point charge at the center. (Although the system exists in three-dimensional space, for ease of plotting we assume everything lies in the plane of the page.)

We begin with the definition of the electric field given in Eq. (19.2):

$$\vec{E}_s \equiv \frac{\vec{F}_{s \to t}}{q_t}$$

In this case the collection of source charges is simply a single point charge, and we replace the subscript "s" with "p" to remind us that we are considering a single point charge:

$$\vec{E}_p \equiv \frac{\vec{F}_{p \to t}}{q_t}$$

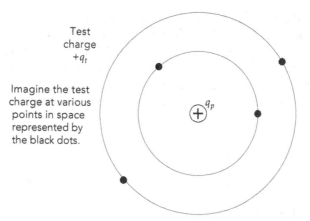

Fig. 19.7. A test charge q_t is located at various positions (black dots) near a positive point charge q_p.

The quantity $\vec{F}_{p \to t}$ represents the force on the test charge due to the point charge at the center. But this is nothing more than the force between two point charges, which we know from Coulomb's law in Eq. (19.1). Plugging in from Coulomb's law with $q_A \to q_p$ (the point charge) and $q_B \to q_t$ (the test charge), we have

$$\vec{E}_p \equiv \frac{k q_p q_t}{r^2 q_t} \hat{r}_{p \to t} = \frac{k q_p}{r^2} \hat{r}_{p \to t}$$

where r is the distance between the point charge at the center and the location of the test charge. Notice that q_t cancels from the equation (this is essentially the reason the electric field is defined the way it is!). We now have something called the electric field that represents a quantity associated only with the collection of source charges (in this case, a single point charge).

The vector $\hat{r}_{p \to t}$ represents a unit vector (length of one) directed from the point charge to location of the test charge. Although the test charge can be placed anywhere in space, the unit vector $\hat{r}_{p \to t}$ will always point *radially outward from the point charge*. Because of this, it is customary to write the unit vector as $\hat{r}_p$ (or simply $\hat{r}$). We therefore arrive at an expression for the electric field due to a single point charge:

$$\vec{E}_p = \frac{k q_p}{r^2} \hat{r}_p \quad \text{or} \quad \vec{E}_p = \frac{k q}{r^2} \hat{r} \quad \text{(for a point charge)} \qquad (19.4)$$

The second form simply eliminates the subscripts on the right side; the point charge is assumed to have charge q and the vector $\hat{r}$ indicates a unit vector pointing radially outward from this charge.

19.6.1. Activity: Electric Field Vectors from a Single Point Charge

a. In the diagram below, sketch some of the electric field vectors due to a *positive* point charge. (While there are an infinite number of locations where you can place the test charge, we will simply choose the four shown!). The length of each vector should indicate the *relative* magnitude of the electric field (if the field is stronger at one point than another, make its vector longer). Of course, the direction of the vector should be in the direction of the field. And finally, the convention is to always place the *tail* of the E-field vector at the point of interest (the test charge), rather than at the location of the charged object generating the field.

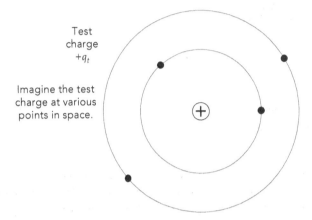

b. Similarly, sketch some of the electric field vectors due to a *negative* point charge. Think about how changing the sign of the charge affects the electric field vector in Eq. 19.4.

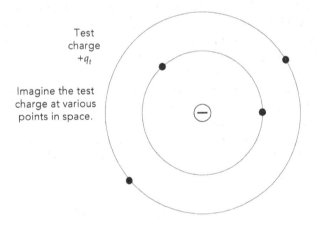

Electric Field of a Collection of Charges

The fact that the electric field is a vector means that the electric fields from a set of charged objects at different points in space can be added together vectorially. Thus, we can always find the total electric field using the *principle of superposition* (we simply add them together). This procedure is discussed in the following activity.

19.6.2. Activity: Electric Field Vectors from Two Point Charges

The diagram below shows two charges: $q_A = + 2.0 \times 10^{-9}$ C at position $(x, y) = (-4 \text{ cm}, 0)$ and charge $q_B = - 2.0 \times 10^{-9}$ C at position $(4 \text{ cm}, 0)$. We wish to calculate the net (or total) electric field due to these two charges.

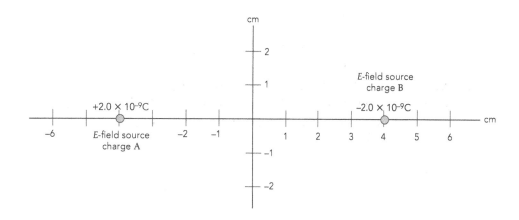

a. Consider the point $(0, 0)$. Determine the electric field at this location due *only* to charge q_A. Include both magnitude and direction.

b. Next, draw a vector representing the electric field from charge q_A at point $(0, 0)$. Put the *tail* of the vector at the point $(0, 0)$ since this is the point of interest. Label this vector $\vec{E}_A$.

c. Repeat this process (calculation and drawing) for the electric field at point $(0, 0)$ due *only* to charge q_B, remembering to put the tail of vector $\vec{E}_B$ at the point $(0, 0)$.

d. Finally, calculate the *net* electric field from both charges q_A and q_B at the point $(0, 0)$. Give the field in vector component form. Based on your result, draw in the *net* electric field vector $\vec{E}_{net}$, putting the *tail* of the vector at the point $(0, 0)$.

e. Now, consider the point (6 cm, 0 cm) instead. Repeat the process above to find the net electric field at this new point.

f. Do the same thing for the point (0 cm, 2 cm). (What do you notice about the y-component of the net field?) After finishing, write down the electric field at the point (0 cm, −2 cm) *without* doing any calculations.

g. Finally, calculate the net electric field at the point (−2 cm, 2 cm).

19.7   THE ELECTRIC FIELD FROM AN EXTENDED CHARGE DISTRIBUTION

We just saw that the electric field from two point charges is simply the sum (superposition) of the fields from the individual charges. This same idea holds true as the number of charges in the collection grows; in principle, you simply add them all up. In practice, however, it quickly becomes unwieldy to be adding up hundreds (or even more) point charges. Instead, it is easier to consider a *continuous charge distribution*, similar to using a mass distribution (or mass density) for determining the center of mass of an object. Here, it is the electric charge density that matters, and as with the center of mass, the situation simplifies further if the charge density is uniform.

Although electric charges will not usually distribute themselves uniformly throughout a conductor, charge can be distributed uniformly throughout an insulator, and we can think about adding up the electric fields from different portions of the object. For example, we can divide the object into small segments, each of which contains a tiny amount of charge Δq_i. Then, by assuming

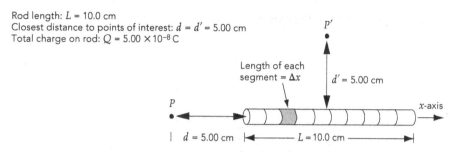

Rod length: $L = 10.0$ cm
Closest distance to points of interest: $d = d' = 5.00$ cm
Total charge on rod: $Q = 5.00 \times 10^{-8}$ C

Fig. 19.8. Diagram of a charged rod broken into ten imaginary "point" charges for a calculation of the electric field at points P and P'.

that each segment behaves like a point charge, the electric field at some point in space can be calculated for each segment. The total electric field at the point of interest is simply the vector sum of the contributions of each of the charge segments.

This process yields an approximate value of the electric field at that point. To get a more accurate value, we could break the object up into more segments. The exact value comes by summing up infinitely many, infinitesimally-small elements, and as we saw with the center of mass, this process leads to an integral.

In this section, we will consider a cylindrical rod with uniform charge distribution throughout. The cylinder is broken up into ten segments, each of which approximates a small point charge. Our goal is to calculate the electric field at two points in space, P and P', as shown in Fig. (19.8). While you could calculate the electric field at any point in space, the symmetry of these two points makes the problem much easier. We begin by qualitatively drawing the field vectors for the individual segments. This is followed by an approximate numerical calculation using a spreadsheet. We will then finish by doing the "exact" calculation using an integral and comparing the two methods of calculation.

19.7.1. Activity: E-Field Vectors from a Uniformly Charged Rod

In each case—point P in part (a) and point P' in part (b)—draw vectors representing the electric fields from the individual segments of the rod. It's probably excessive to draw the vectors for all ten pieces—we would suggest drawing four field vectors, corresponding to segments 1, 2, 9, and 10. The length of the vectors should correspond to the approximate *relative* magnitudes of the different contributions. Remember, it is customary to locate the *tail* of each E-field vector at point of interest (either P or P').

a. Parallel to the axis of the rod

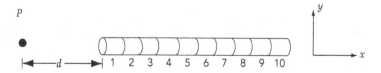

b. Perpendicular to the axis of the rod

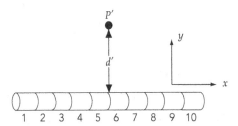

In part (a) all the field vectors point in the same direction (to the left, or the negative x-direction). This makes the problem *much* easier, because the total (or net) electric field will simply be a larger vector pointing in the same direction. In part (b), the x components of the individual field vectors cancel when they are added. This is because point P' is at the exact midpoint of the rod, so the (positive) x-component of $\vec{E}_1$ cancels with the (negative) x-component of $\vec{E}_{10}$. The same thing occurs for segments 2 and 9, etc. Once again, this symmetry leads to a *tremendous* simplification; because the x-components cancel, the net electric field will be a vector pointing only in the $+\hat{y}$ direction. Thus, we only need to worry about adding together the y-components when calculating the electric field at point P'.

19.7.2. Activity: Approximate Field Calculation Along the Axis of a Rod

a. Consider the case of point P, along the end of the rod. Assuming the rod is divided into ten segments, determine the amount of charge Δq_i contained in each segment in terms of the total charge Q. Remember that the charge is assumed to be uniformly, or evenly, spread throughout. (This is straightforward—don't overanalyze!)

When considering discrete segments, the net field is determined by summing up the individual contributions, each of which is considered a point charge:

$$\vec{E}^{\text{tot}} \approx \sum_i \vec{E}_i = \sum_i \frac{k \Delta q_i}{r_i^2} \hat{r}_i$$

For our situation of point P along the axis, all the vectors point in the same direction, so we can drop the vector notation (for now) and worry only about the magnitude. Notice what a big simplification this is; instead of adding vectors, we only need to add numbers! We can also put everything in terms of the variable x, since both point P and the rod lie along the x-axis:

$$E^{\text{tot}} \approx \sum_i \frac{k \Delta q_i}{x_i^2}$$

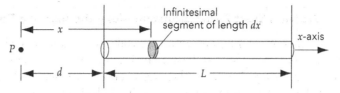

Fig. 19.9. Diagram showing an infinitesimal length dx at an arbitrary distance along the rod of length L. The distance from point P to the end of the rod is d, while the distance to the given infinitesimal segment is denoted as x.

where k is Coulomb's constant, Δq_i is the charge of each segment, and x_i is the distance of segment i from point P. (You will need to determine an expression for x_i in part (b).)

b. Using a spreadsheet (or other program), calculate the magnitude of the electric field at point P for each of the ten segments (you found Δq_i in part (a), but you still need to find x_i). Remember, we are assuming that each segment is small enough that it can be treated as a point charge *located at the center of each segment*. Once you have the electric field for each segment, sum up the ten values to find E_{net}. Finally, write down your final answer for $\vec{E}_{\text{net}}$ in vector notation below (and affix a printout of your spreadsheet if your instructor requests).

The calculation you just completed provides a numerical approximation to the electric field on the axis of the rod for our specific situation. (It is only approximate since each segment has some finite width, and so we can't really assume all of the charge in that segment is located at its center). It is possible to use integration to derive a general equation giving the magnitude of the electric field at any point P along the axis of the rod. To keep things general, we will write everything in terms of the variables Q (total charge), d (distance from the end of rod), and L (rod length).

For the exact calculation, this sum becomes an integral with the appropriate substitutions:

$$E^{\text{tot}} = \sum_i \frac{k\Delta q_i}{x_i^2} \rightarrow \int_{x_{\min}}^{x_{\max}} \frac{k\, dq}{x^2}$$

where x represents the distance from point P to the infinitesimal segment dx, dq is the infinitesimal amount of charge in this segment, and the integral is performed over the length of the rod (see Fig. 19.9).

19.7.3. Activity: Exact Field Calculation Along the Axis of a Rod

a. Notice that as we integrate along the rod, the variable x is changing, so we should expect there to be a dx inside the integral. There is, but it's hidden in the dq. Use the total charge Q and the rod length L to relate dq to dx. **Hint**: If a length of rod L contains a charge Q, how much charge is contained in a segment dx?

b. Based on the diagram, determine the values of x_{min} and x_{max}. Remember, the integral should go from one end of the rod to the other.

c. Substitute your expressions for dq, x_{min}, and x_{max} into the integral and perform the integration (don't forget to evaluate it at the end points). What is the final expression for the magnitude of the total electric field?

d. You should have found that $E^{tot} = \frac{kQ}{d(L+d)}$. So that we can compare with the approximate solution, substitute in numerical values for the variables Q, d, and L (and k) from Fig. 19.8. How does the approximate value you found in Activity 19.7.2 compare to the exact value you just calculated? Compute the percent discrepancy. How could you make the numerical method more exact?

19.8 PROBLEM SOLVING

Imagine that your physics lab contains a collection of small balloons that have been (uniformly) charged by rubbing them with either polyester or fur. You place balloon A with charge $q_A = +1.0 \times 10^{-6}$ C at one end of your table. For simplicity, we'll locate the point (0 m, 0 m), or the origin our of xy-coordinate system, at the center of balloon A. Balloon B has charge $q_B = +3.0 \times 10^{-6}$ C and is placed on your table with its center at the point (0.40 m, 0.30 m).

19.8.1. Activity: Two-Dimensional Fields

a. Draw a diagram with axes showing the locations of the balloons. Assume the balloons are small compared to their separation so that they can be considered point charges.

b. We are interested in calculating the electric field at the point $P = (0.20$ m, -0.20 m) on your table so that it can be compared to a subsequent experimental measurement. Start by sketching two vectors that show the electric field due to each balloon separately at point P. Label the vectors $\vec{E}_A$ and $\vec{E}_B$, being sure to locate the tail of each vector at point P and that the relative lengths of the vectors are correct.

c. Find the net (total) electric field $\vec{E}_{net}$ at the point P. Write the net field using *both* vector components *and* a magnitude with direction (angle).

d. Now imagine placing a third balloon (balloon C) with charge $q_C = -4.0 \times 10^{-6}$ C at point P. What is the net electric *force* on balloon C due to the presence of balloons A and B? Write the force using *either* vector components *or* a magnitude with direction. **Hint**: Do it the quick way; you do not need to do any more vector superposition!

UNIT 20: ELECTRIC FLUX AND GAUSS'S LAW

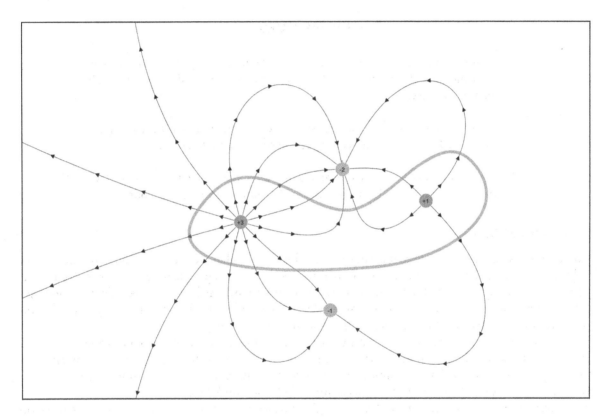

A collection of positive and negative charges are placed at different locations, resulting in a complicated pattern of electric field lines. If we draw a simple closed path (as shown by the thick curve), it turns out that the electric flux through this "surface" is related to the total charge enclosed. This is the essence of Gauss's law, and once you complete this unit, you should be able to use this law to determine the electric field from a symmetric charge distribution.

UNIT 20: ELECTRIC FLUX AND GAUSS'S LAW

OBJECTIVES

1. To understand how electric field lines can be used to represent the electric field and how electric flux relates to how the electric field passes through an area of space.

2. To discover the relationship between the electric flux passing through a closed surface and the net charge enclosed by that surface.

3. To explore the concept of symmetry and use Gauss's law to calculate electric fields that result from symmetric distributions of electric charge.

20.1 OVERVIEW

Coulomb's law and the principle of superposition can be used to calculate the force (and hence the electric field) on a charge due to distributions of "source charges." It is also possible to calculate the electric field using a completely different formulation known as Gauss's law, which relates the field surrounding a collection of charges to the amount of charge enclosed by a surface. Gauss's law is a very powerful tool for calculating electric fields due to *symmetric* distributions of charge.

We will begin the study of Gauss's law by learning about a convenient construct known as *electric field lines*. These lines can be used to map the direction of the electric field, and therefore the direction of the net force on a test charge, at any point in space. We will practice constructing electric field lines from a configuration of charges using superposition and Coulomb's law.

Next, we will consider the idea of *flux* through a surface and how this relates to electric field vectors and lines. We will examine "closed surfaces" around various charges or groups of charges and see how many electric field lines pass in and out of the surfaces. Finally, we will explore the concept of symmetry and use the mathematical representation of Gauss's law to calculate the electric field at various points in space due to different charge distributions.

ELECTRIC FIELD LINES AND FLUX

20.2   ELECTRIC FIELD LINES

We have been representing the electric field due to a configuration of charges by a vector that indicates both magnitude and direction. These electric field vectors exist at every point in space; at any given point, you could find both the magnitude and direction of the electric field and, in general, the values will be different everywhere. This is the conventional representation of what is known as a *vector field*.

As you can probably imagine, this description can get unwieldy, as there are vectors of varying length and direction everywhere. An alternative representation of the electric field involves defining *electric field lines*. Unlike an electric field vector, which is an arrow with magnitude and direction, electric field lines are continuous and generally simpler to construct. The two representations (vector fields and field lines) are complementary, and one representation may work better than the other, depending on the situation.

In the following activity, you will explore some of the properties of electric field lines for simple situations. Your instructor will provide you with a method of generating electric field vectors and lines for different situations.[1]

20.2.1. Activity: Simulation of Electric Field Lines from Point Charges

a. Place a single charge of +1 (we'll assume the units are Coulombs) in the region. Indicate the charge by writing +1 next to the charge. You should see a set of field lines emanating from the charge. Sketch the lines in the space below, being sure to show the direction of each line by placing arrows on them. How many lines are there in the drawing? Are the lines more dense or less dense near the charge?

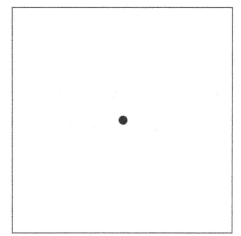

b. To understand how electric field lines differ from electric field vectors, choose a couple of locations in the diagram above and draw in the *vector*

---

[1] One electronic resource is the Electric Field Line Simulator from academo: https://academo.org/demos/electric-field-line-simulator/

representing the electric field at each location. Be sure to choose locations at different distances from the charge and on different "sides" of
the charge.

c. Now change the value to try both +2 units of charge and −1 units of
charge. No need to sketch the results, but be sure to comment on how
many lines there are, as well as the directions of the lines. Can you
describe the rule for determining how many lines emanate from a particular amount of charge?[2] How does the direction of the lines depend
on the sign of the charge?

d. Repeat the exercise using two charges with the same magnitude but having unlike signs. Place them at two different locations as shown below.
Sketch the field lines and indicate how much charge is in each location.
Do the rules you determined in part (c) hold? Where are the lines most
dense?

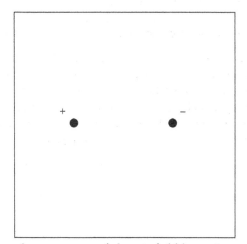

e. Summarize the properties of electric field lines. In particular, what does
the *number* of lines emanating from a charge signify? What does the
direction of a line represent? What does the *density* of the lines reveal?

---

[2] The actual number you get depends on the settings in the simulation used to generate
the diagram. There is no universally-accepted number, and one clearly can't have one
line per Coulomb of charge, as you would be at a loss as to how to draw lines for 10^{-6}
Coulombs (how do you draw 10^{-6} lines?). All we can say is that the number of field lines
is *proportional* to the charge, and that the constant of proportionality, while arbitrary,
should remain fixed when comparing different situations.

20.3 ELECTRIC FLUX

As we've just seen, we can think of an electric charge as having a specific number of electric field lines (either converging on it or diverging from it) that is proportional to the magnitude of the charge. These lines are a representation of the electric field that exists in the space surrounding the charge. *Electric flux* is defined as the number of electric field lines passing through a surface. In defining the concept of "flux," we are constructing a mental model of electric field lines streaming out of (or into) each unit of charge like streams of water or rays of light. These streams might cross through a surface located somewhere in space. For example, consider rays of light coming from the sun streaming through a window. Depending on the size of the window and the angle of the sun, a certain number of these rays will pass through the window (they will "pierce the surface"), which represents the flux through the window surface.

The same concept holds true for electric field lines passing through an arbitrary surface. This surface could be an actual, physical surface (like a window), or it could be imaginary—something we construct in our minds. While we obviously can't see electric field lines streaming out from charges, the conceptual description for charges and fields is the same as for streams of water or rays of light.

Hopefully, it's clear that the number of field lines passing through a surface depends not only on the area of the surface, but also on how that surface is oriented relative to the field lines. The setting sun often shines directly into a window, while the overhead sun at noon may not pass through the window at all, instead shining downward "parallel" to the glass. The size and orientation of a surface of area A is usually given as a vector that is perpendicular ("normal") to the surface and has a magnitude equal to the surface area. By convention, this is called the *normal vector to the surface* and points away from the outside of the surface.

Figure 20.1 shows three different surfaces with area vectors. For an infinitesimally thin surface such as a plane (left image), there are two possible directions of $\vec{A}$: "up" or "down" from the surface. You are free to choose which way $\vec{A}$ should point, but once the choice is made, you need to stick with it for the duration. For a closed surface such as a cubical box (middle image), convention says that the area vector $\vec{A}$ points *out* from the surface, *away* from the enclosed volume.[3] Note that for a six-sided cube, there are actually six

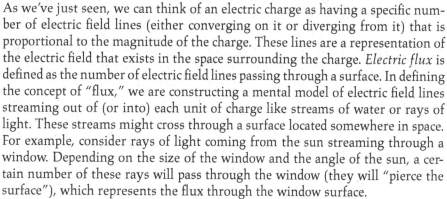

Fig. 20.1. Diagrams showing how a normal vector representing the area points away from the surface.

---

[3] As we will see, this choice means that electric field lines passing from the inside to the outside of a closed surface lead to positive flux, while field lines passing from the outside to the inside of a surface lead to negative flux.

possible area vectors (one for each surface). Each area vector will point outward from the enclosed volume, and all vectors will have the same length if the sides have equal areas. For a curved surface such as the outer part of a cylinder (right image), we will want to think of (infinitely) many small area element vectors $d\vec{A}$ that can be located anywhere on the curved surface and always point outward from the enclosed volume. The top (and bottom) surfaces of the cylinder are flat and can be represented with standard area vectors.

We are interested in determining how the flux through a surface depends on the angle between the area vector and the electric field lines. To answer this question, we will use a mechanical model of electric field lines and a surface. As one example, you can arrange nails in a 10 × 10 array poking up at 1/4″ intervals through a piece of wood or Styrofoam. The surface can be a flat region (a plane) bounded by a square wire loop. If you wish to construct this model, you will need:

- Styrofoam or 3/8″ plywood (5″ × 5″ square)
- 100 nails, approximately 4″ in length (mounted on the Styrofoam or plywood)
- 1 wire loop (4″ × 4″ square)
- 1 paper, 5″ × 5″ with 1/4″ graph rulings (to affix to the mounting square to help with spacing the nails)
- 1 protractor

Figure 20.2 shows a rendering of a board with a square array of nails, along with a square wire loop. The nails represent electric field lines, which are assumed to point in the direction of the nails (directly upward from the base). The square wire loop represents the boundary of a planar surface with its area vector shown; we have located the area vector along the closest edge of the surface (instead of the traditional center of the surface) so that it can be

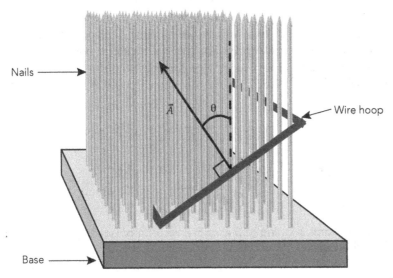

Fig. 20.2. Apparatus to determine how many uniformly-spaced field lines will pass through an imaginary surface area as a function of the angle between the direction of the field lines and the normal vector representing the surface area.

easily seen. The wire loop can be adjusted to have a known number of nails pass through its surface. Use the fact that the nails are arranged in a 10 × 10 grid to determine the number of nails passing through the loop without actually counting them!

20.3.1. Activity: Flux as a Function of Angle

a. Use your mechanical model and a protractor to fill in the following data table (you may want to put the data directly into graphing software, as we will be plotting the data in a moment). Orient the wire loop to contain the specific number of lines given in the table below, and then use the protractor to determine the angle θ between the electric field lines (the nails) and the normal vector to the surface. You might need to think carefully about how to see this angle on the protractor! Negative values of flux correspond to times when the wire loop has been "flipped over" so that its area vector is pointing below the horizontal (the nails point the "other way" through the loop). **Hint**: It's possible to use symmetry to fill in the second half of the table without having to make any more measurements!

Flux Φ (number of lines piercing the surface)	Angle θ between electric field and area vector (degrees or radians)
100	
90	
80	
70	
60	
50	
40	
30	
20	
10	
0	
−10	
−20	
−30	
−40	
−50	
−60	
−70	
−80	
−90	
−100	

b. Use a graphing program to plot the flux Φ as a function of the angle θ. This graph should look like a curve instead of a straight line. Can you guess a mathematical function that might describe this curve? (This is a bit challenging, since we are only seeing a portion of the curve).

c. Try fitting a cosine curve to your data, adjusting the amplitude and frequency to best match the data. How does the fit look? Print out or sketch the plot (including the mathematical fit) in the space below.

d. Remember the dot product for two vectors? We saw that we could write the dot product of two vectors in terms of vector magnitudes and the angle between them: $\vec{a} \cdot \vec{b} = |\vec{a}| \, |\vec{b}| \cos \theta$, where θ is the angle between vectors $\vec{a}$ and $\vec{b}$. Based on your fit and this expression for the dot product, write down an equation that gives the electric flux through a surface in terms of the vectors $\vec{E}$ (the electric field) and $\vec{A}$ (the area vector for the surface), assuming the electric field is uniform over the surface. Use the variable Φ for the flux.

Defining Flux

Qualitatively, we said that flux is related to the number of field lines that pierce through a surface. We quantified this in the above activity for the case of a uniform electric field and saw that the flux could be represented by the vector dot product between the electric field and the area vector:

$$\Phi = \vec{E} \cdot \vec{A} \quad \left(\text{flux for uniform electric field through a surface}\right) \qquad (20.1)$$

Note that flux is a scalar quantity (*not* a vector). In general, the electric field will not be uniform, and the surface will not be a simple plane. In such a situation we must break the surface into tiny (infinitesimal) areas $d\vec{A}$ that are approximately flat (e.g., see Fig. 20.1). Then the flux of any infinitesimally small patch of surface will be given by $d\Phi = \vec{E} \cdot d\vec{A}$, where $\vec{E}$ is the (vector) electric field over that small patch (assumed to be uniform since the area of the patch is infinitesimally small). The net flux is then calculated by adding up (integrating) all the flux elements:

$$\Phi^{net} = \int d\Phi = \int \vec{E} \cdot d\vec{A} \quad \left(\text{net flux through a surface}\right)$$

Some surfaces, like that of a sphere or the six surfaces that make up a rectangular box, are closed surfaces. A *closed surface* has no holes, so nothing can leave its interior without passing through the surface. There is a special notation to represent the integral over a closed surface, which is denoted as follows:

$$\Phi^{net} = \oint d\Phi = \oint \vec{E} \cdot d\vec{A} \quad \text{(net flux through a closed surface)} \qquad (20.2)$$

As we will see, it is the flux passing through a closed surface that will be relevant for Gauss's law.

20.4 GAUSS'S LAW

We are interested in determining how the flux passing through a closed surface is related to the charges that generate the electric field. It is easiest if we start by considering a two-dimensional world, in which all charges and electric field lines are constrained to lie in a flat, two-dimensional space (a plane).[4] In two dimensions, "closed surfaces" are nothing more than simple closed curves (or lines) that lie in a given plane. For example, circles, squares, ovals, etc., are all closed "surfaces" in two dimensions, as is any arbitrarily shaped curve that eventually ends up back where it started. In the activity below, we examine how the net number of field lines passing through a surface is related to the net charge enclosed by the surface. In essence, this relationship is what we refer to as Gauss's Law.

---

[4] A two-dimensional world would be a strange place to live! E. A. Abbot's book entitled *Flatland; A Romance of Many Dimensions* (Dover, New York, 1952) considers such a world. It's a delightful piece of late nineteenth-century political satire in the guise of a mathematical spoof.

20.4.1. Activity: Gauss's Law in Two Dimensions

a. Figure 20.3 shows a configuration of charges and the associated electric field lines over some (two-dimensional) region. Draw in four different, two-dimensional closed "surfaces" in the region. Each of your surfaces can have any shape you want, but they should not extend beyond the rectangular box containing the image. Try to have each surface contain a different number of charges. Count the net number of lines piercing each "surface" to determine the net flux. **Note**: Lines coming *out* of a surface are considered to be positive, while lines going *into* a surface are considered to be negative. The *net flux* (in units of field lines) is then simply the number of lines coming out of the surface minus the number of lines going into the surface. Use this information to fill in the table below.

	Charge enclosed by the arbitrary surface			Lines of flux out of and into the surface		
Surface	Total Positive Charge (Arbitrary Units)	Total Negative Charge (Arbitrary Units)	Net Charge Enclosed	Φ^{out}	Φ^{in}	Φ^{net}
1						
2						
3						
4						

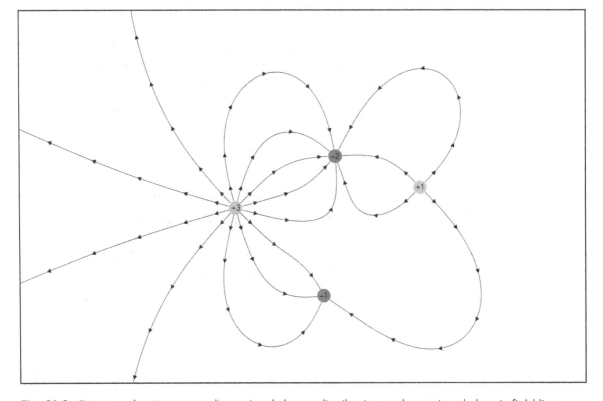

Fig. 20.3. Diagram showing a two-dimensional charge distribution and associated electric field lines.

b. Based on the table, can you find a relationship between the net flux and the net charge enclosed for all your (closed) surfaces?

You should have found that the net flux through any closed surface was *proportional* to the net charge enclosed. For this particular diagram the constant of proportionality was four (with units of field lines per charge). Note that this number is arbitrary—when the diagram was created, a choice was made for how many field lines to draw per unit of charge. In a real situation, this constant of proportionality would have real units that relate flux and charge. We will see the "real" value of this constant of proportionality in the next section.

Gauss's Law in Three Dimensions

We just considered a situation in two spatial dimensions. In reality, of course, we live in three spatial dimensions, and so we should consider closed surfaces in three dimensions. If we do this, we find that the same relationship holds. In words, the net flux through a closed surface is proportional to the net charge enclosed by the surface. Mathematically, this statement is given by

$$\Phi^{\text{net}} = \oint \vec{E} \cdot d\vec{A} \propto q_{\text{encl}} \tag{20.3}$$

where q_{encl} is the *net* charge enclosed by the surface.

GAUSS'S LAW, ELECTRIC FIELDS, AND CONDUCTORS

20.5   USING GAUSS'S LAW TO CALCULATE ELECTRIC FIELDS

In SI units the constant of proportionality in Eq. (20.3) ends up being $4\pi k$, where k is Coulomb's constant (from Coulomb's law). Equation (20.3) is sometimes written in terms of a different constant known as the electric permittivity, which in vacuum has a value of $\varepsilon_0 = 8.85 \times 10^{-12}$ C^2/N m^2. These two constants are related by $\varepsilon_0 = 1/(4\pi k)$, and Gauss's law can be written mathematically using either as[5]

$$\Phi^{\text{net}} = \oint \vec{E} \cdot d\vec{A} = \frac{q_{\text{encl}}}{\varepsilon_0} = 4\pi k \, q_{\text{encl}} \quad \text{(Gauss's Law)} \tag{20.4}$$

---

[5] This expression is a complicated mathematical beast involving three-dimensional vector calculus! Don't be worried if you have never seen anything like this before; we will work our way through each part step by step.

where Φ^{net} is the net flux through any arbitrary closed surface, and q_{encl} is the net charge enclosed by that surface. Gauss's law has a variety of applications, one of which is to compute the electric field at some point in space due to a distribution of charges. While Gauss's law is always true, it turns out to be most useful in situations that involve a *symmetric* charge distribution. What is a symmetric charge distribution? It's an arrangement of charges that can be rotated about an axis and/or reflected in a mirror and still look the same.

For example, consider a sphere that has no pattern or design on the outside. This sphere looks the same from any angle; it doesn't matter how you rotate it or where you stand, it always looks the same. This object has a high degree of symmetry (a *spherical symmetry*). However, if you put some markings on the sphere (e.g., a basketball or a soccer ball), the ball will look different when rotated slightly, as the pattern will shift.[6] Similarly, a perfect cylinder has *cylindrical symmetry*—you can rotate the cylinder around the axis of the cylinder by any amount, and it will still look the same. Note, however, that this is not true if you rotate a cylinder about a different axis. As we'll see, the key is to pick a closed surface with the same symmetry as the electric field associated with the charge distribution.

Learning to apply Gauss's law will take some practice, and so we start with an activity that guides you through the steps. In fact, we'll begin with an example for which we already know the answer—that of a point charge!

20.5.1. Activity: Electric Field from a Point Charge Using Gauss's Law

We will use Gauss's law to calculate the electric field at any distance away from a single point charge. As noted, we already found the solution in Section 19.6 (see Eq. (19.4)). But we will use this activity to outline the steps for applying Gauss's law. The main steps are summarized after we talk through them; hopefully you will find these helpful when applying Gauss's law to other situations. Please read through the procedure as you answer the questions, discussing them with your partners to make sure everyone understands each step (it is subtle!)

Figure 20.4 shows a positive (point) charge q_0 with electric field lines sketched in. We were able to draw the field lines based on our past experience. We begin with the equation for Gauss's law:

$$\oint \vec{E} \cdot d\vec{A} = \frac{q_{encl}}{\varepsilon_0}$$

---

[6] Of course, if you rotate the ball a certain amount in a particular direction, the repeating pattern may line itself back up, and once again look the same. The amount the ball must rotate for this to occur depends on the details of the pattern, and there is a whole subbranch of physics and math that deals with these types of symmetries (we will not concern ourselves with these subtleties).

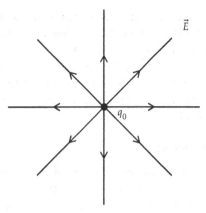

Fig. 20.4. A point charge q_0 (assumed to be positive) with electric fields lines emanating outward.

Here, the left side represents the net flux through a closed surface (*any* closed surface), while the right side includes the net charge enclosed by this surface. We will consider the left side and right side of this equation separately, starting with the left, which involves the dot product of the electric field with the area vector over the closed surface. As in Activity 20.4.1, we are free to choose any closed surface we want, but it's important that the symmetry of the surface match the symmetry of the charge distribution.

Step 1: Draw the Gaussian Surface

The charge distribution in this example consists of a single point charge, assumed to be infinitesimally small. This point charge looks the same from any direction, so this situation has *spherical symmetry* (you can think of the point charge as a tiny sphere with infinitesimally small radius). Thus, the imaginary surface we draw should have the same symmetry, so we choose a spherical surface (a *shell*) having an arbitrary radius r. This surface is sketched in Fig. 20.5 (remember, we are only drawing a surface, not a solid sphere!).

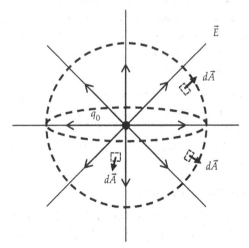

Fig. 20.5. An imaginary spherical shell (a Gaussian surface) surrounds a point charge. Area element vectors $d\vec{A}$ are always normal to the surface.

Step 1 Summary: Draw an imaginary "Gaussian surface" with the *appropriate spherical symmetry* that surrounds the charge. Technically, this is a three-dimensional situation, and so the diagram attempts to show a three-dimensional spherical shell (one can often get away with drawing everything in the two dimensions of the page, but it might require multiple views in some situations). It is customary to use a dashed line to represent the Gaussian surface.

Step 2: Draw the Vectors

The integral on the left side involves the dot product of the electric field vector $\vec{E}$ and the area element vector $d\vec{A}$. Because we are dealing with a closed surface integral, we need to *evaluate the dot product at all points on the surface*. In other words, at every point on our imaginary Gaussian surface we need to look at both $\vec{E}$ and $d\vec{A}$. Fortunately, the symmetry of the situation makes this feasible!

a. At any point on our imaginary surface, describe the orientation of the electric field vector $\vec{E}$. In other words, what direction does this vector point?

b. Similarly, at any point on our imaginary surface, describe the orientation of the vector $d\vec{A}$. In other words, what direction does the vector $d\vec{A}$ point?

You should have concluded that at *any* point on the Gaussian surface, both $\vec{E}$ and $d\vec{A}$ point *radially outward* (away from the charge). Note that you can't simply say the x-direction (or z-direction, etc.), as both $\vec{E}$ and $d\vec{A}$ point in different directions as you move around the imaginary surface. But they always point in the *same* direction at any given point!

Step 2 Summary: Draw vectors representing both $\vec{E}$ and $d\vec{A}$ at different points on the imaginary Gaussian surface using the symmetry of the situation to do so. It is beneficial to draw the vectors at a few different places on the surface, especially in situations where the surface has different faces.

Step 3: Evaluate the Dot Product

We are now ready to consider the integral on the left side, which involves the dot product $\vec{E} \cdot d\vec{A}$ at every point on the Gaussian surface. We know we can write the dot product as

$$\vec{E} \cdot d\vec{A} = |\vec{E}||d\vec{A}| \cos\theta = E\, dA \cos\theta$$

where θ is the angle between the two vectors.

    c. Given your answers to parts (a) and (b), what is the angle between the vectors $\vec{E}$ and $d\vec{A}$ at *any* point on the imaginary Gaussian surface? Use this fact to calculate the dot product $\vec{E} \cdot d\vec{A}$.

Because the vectors $\vec{E}$ and $d\vec{A}$ point in the same direction (radially outward) at every point on the Gaussian surface, the angle between these two vectors is always zero: $\theta = 0$ at every point on the surface. Note the symmetry is critical here; if one vector was changing while the other was not, we could not make this conclusion. While the angle will not always be zero degrees in every problem, we will find that symmetry usually dictates that $\theta = 0$, 90°, or 180°. If the surface has different "faces" (e.g., a cylinder with ends as opposed to a sphere), the angle will likely have different values on the different faces. In such a situation we must consider the integral over each face separately. (Note that for any face where $\theta = 90°$, the dot product yields zero so there will be no contribution to the integral from this face.)
For the current situation, the integral on the left side is reduced to

$$\oint \vec{E} \cdot d\vec{A} = \oint E \, dA$$

where $E = |\vec{E}|$ is the *magnitude* of the field on the surface (similarly for dA).

Step 3 Summary: Determine the angle between vectors $\vec{E}$ and $d\vec{A}$ at all points on the Gaussian surface. Use this angle to evaluate the dot product. If the surface has different faces, there may be different values of the angle on the various faces (if so, proceed with Step 3a).

Step 3a Summary: If the angle between vectors $\vec{E}$ and $d\vec{A}$ varies over different faces of the surface, split up the integral into different pieces, with each piece representing the integral over a particular face of the surface.

Step 4: Move E Out of the Integral

Since we don't know the value of E (this is what we are trying to find after all!), we would like to move it out of the integral to proceed. However, we can only do this if E is constant over the entire region of the integral.

    d. Explain why the magnitude of the electric field is the same at all points on our Gaussian surface. **Hint**: How does E depend on the distance from the charge?

Our Gaussian surface is a spherical shell that is *always the same distance away from the charge*. Therefore, by symmetry we know that the magnitude of the electric field must be the same everywhere on the surface. Alternatively, one can think about Coulomb's law and putting a test charge at any point on the Gaussian surface; because the distance from the test charge to the charge at the center is always the same (it's the radius r of the shell), the force on the test charge will be the same everywhere on the surface. The result is that the magnitude of the electric field is also the same everywhere on the surface.

It's important to realize that we can only draw this conclusion because the symmetry of the Gaussian surface matches the symmetry of the charge distribution (and therefore the electric field). For example, if we had drawn a cubical shell instead of a spherical shell for our Gaussian surface, not only would the angle between $\vec{E}$ and $d\vec{A}$ change as you move around the surface, but the magnitude of the field would be weaker at the corners where you are farther away from the charge. Only by drawing a Gaussian surface that mimics the symmetry of the charge distribution (a single point charge in this case) are we able to conclude that the field has the same magnitude everywhere on the surface. This is an important logical step—please make sure you understand it before moving on!

Since E has the same value everywhere on the surface, it is constant, and we can pull it out of the integral:

$$\oint \vec{E} \cdot d\vec{A} = \oint E \, dA = E \oint dA \ \left(\text{since } E \text{ constant over surface}\right)$$

Step 4 Summary: Verify that, due to symmetry, the magnitude of the electric field is constant over the entire surface. Use this fact to pull E out of the integral. (If the surface has different faces and you split up the integral in Step 3a, you should verify that for all faces where $\theta \neq 90°$, the magnitude of the electric field is constant over each individual face.)

Step 5: Evaluate the Integral

On the left side, we are left with evaluating $\oint dA$ (or $\int dA$ over the various remaining faces if the integral has been split up). Fortunately, we know how to evaluate this integral! Just like $\int dx = x$, we know that $\oint dA = A$, where A is the surface area of the closed surface.

e. For our situation, the imaginary Gaussian surface is a spherical shell with arbitrary radius r. What is the surface area of this sphere?

Hopefully you remembered that the surface area of a sphere of radius r is $4\pi r^2$. We have now fully-evaluated left side of Gauss's law:

$$\oint \vec{E} \cdot d\vec{A} = \oint E \, dA = E \oint dA = E \, A = E \, 4\pi r^2$$

Step 5 Summary: Determine A, the surface area of the closed surface. (If the surface has different faces and you split up the integral in Step 3a, you should determine the area of each individual face for which $\theta \neq 90°$ and add these together to get the total surface area.)

Step 6: Determine the Charge Enclosed

Having calculated the left side of Gauss's law, we now turn our attention to the right side, which requires that we determine the charge enclosed by the imaginary Gaussian surface. For this example, finding q_{encl} is simple because there is only one charge in the entire problem and it sits inside our Gaussian surface. Therefore, $q_{encl} = + q_0$! In general, if there are multiple charges or a charge distribution, it will take a little more effort to determine the value of q_{encl}.

Step 6 Summary: Determine q_{encl}, the net charge enclosed by the Gaussian surface. Remember, this is the total charge that is *inside* (enclosed by) the imaginary surface.

Step 7: Solve for the Electric Field

The last step is to set the left side equal to the right side and solve for the quantity of interest. This will typically be the magnitude of the electric field E, but one could imagine a situation where the electric field is known (e.g., it has been measured), and we are interested in determining the enclosed charge instead.

In our example of a single point charge, we have:

$$E\, 4\pi r^2 = \frac{q_0}{\varepsilon_0}$$

Solving for the magnitude of the electric field yields:

$$E = \frac{1}{4\pi r^2}\frac{q_0}{\varepsilon_0} = \frac{1}{4\pi\varepsilon_0}\frac{q_0}{r^2} = \frac{kq_0}{r^2}$$

Back in the initial step, we determined from symmetry that the electric field must point radially outward from the point charge, so the electric field in vector notation for a point charge with charge q_0 is given by:

$$E_p = \frac{kq_0}{r^2}\hat{r} \tag{20.5}$$

where r is the distance away from the charge and $\hat{r}$ is a unit vector that points radially away from the charge. We have put the subscript "p" on E as a reminder that this expression is only valid for a point charge.

f. Compare our result in Eq. (20.5) with what we determined for the same situation using Coulomb's law in Eq. (19.4). Are they the same?

Step 7 Summary: Set the left side equal to the right side and solve for the quantity of interest. If solving for the electric field, be sure to put in the direction of the field using the appropriate unit vector(s).

As expected, the two results for the electric field from a point charge are identical, except for what we chose to label the point charge in the different methods (q versus q_0)! Clearly, using Gauss's law was not necessary for such a simple situation of a point charge; we simply chose this as an example to demonstrate the steps. We will soon encounter more complicated charge distributions that take advantage of the power of Gauss's law.

20.6 ELECTRIC FIELDS AND CHARGES ON CONDUCTORS

An electrical conductor is a material where a subset of the electrons are free to move. Thus, if a free charge in a conductor experiences an electric field, it will move under the influence of that field (in an insulator, charges are bound to their host atoms/molecules and cannot move in the same way). Imagine applying an electric field $\vec{E}_{applied}$ to a region of space and then putting a conductor into the region (see Fig. 20.6). The free electrons will experience a force and move in response to this applied field.

In the situation depicted, the electrons experience a force $\vec{F} = q\vec{E}$ that's opposite the direction of the applied field. The result is a net negative charge near the bottom edge (which leaves a net positive charge near the top edge). Thus, the applied electric field leads to an *induced charge separation* in the conductor. As we know from Unit 19, this means there will also be an electric field created by these induced charges, and we call this induced field $\vec{E}_{induced}$.

Notice that the induced field acts in the opposite direction of the applied field, tending to cancel it out. As long as there is a net field present in the conductor, charges will continue to experience a force. In fact, the charges won't stop moving until there is *no* net field present inside the conductor: $\vec{E}_{induced} = -\vec{E}_{applied}$ so that $\vec{E}_{net} = 0$.

Thus, we can conclude that if there are no moving charges inside a conductor, the electric field inside the conductor must be zero. This is an important result! Simply because the charges are free to move inside a conductor, we are able to conclude that $\vec{E} = 0$ inside a conductor whenever there are no moving charges present (a *static* situation)[7]:

$$\vec{E}_{\text{inside conductor}} = 0 \quad (\text{in static equlibrium}) \tag{20.6}$$

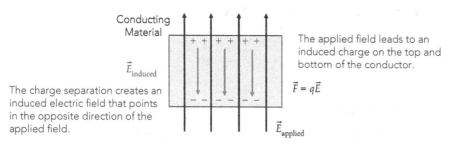

Fig. 20.6. An applied electric field in a conductor results in a charge separation and an induced electric field.

---

[7] Technically this is only true in an *ideal* conductor.

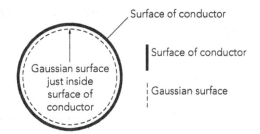

Fig. 20.7. Diagram showing a solid conducting sphere with an imaginary Gaussian surface (also spherical) drawn just inside the surface of the conductor.

In the remainder of this section we look at what additional information this result can tell us about charges and conductors. To perform the experiments in this section you will need the following:

- 1 empty, metal soup can with label removed (or similarly shaped metal object)
- 1 plastic rod (or other source of excess charge)
- 1 piece of fur
- 1 low-mass, conducting ball (like a pith ball) on non-conducting thread
- 1 electroscope (optional)

Excess Charges on a Conductor

We are interested in determining how excess charges arrange themselves on conductors. Consider a *solid* conductor in the shape of a sphere, as shown in Fig. 20.7. After rubbing a plastic rod with fur, you touch the rod to the conductor, which transfer some of the negative charges on the rod to the conducting sphere. Where does this excess charge go if it is free to move through the conductor? Is it distributed uniformly throughout the conductor? Or does it reside somewhere else? To answer this question we will use Gauss's law, making use of the spherical symmetry of the situation.

20.6.1. Activity: Where Is the Excess Charge in a Conductor?

a. Assume you have a solid conducting sphere with a net charge (we'll assume it is negative, but the sign is not important). Now consider drawing an imaginary Gaussian surface *just inside* the outer surface of the conductor such that the Gaussian surface is inside the metal, but just barely. We just found that if the charges are not moving (which is certainly the case if you wait a little while), the electric field inside the conductor is zero. Use this fact, along with Gauss's law, to determine the net charge enclosed inside the Gaussian surface. Explain briefly. **Hint**: You don't need to do any integrals here!

b. This result holds as long as the Gaussian surface is inside the conductor. But if we imagine the surface is just barely inside the material, then according to Gauss's law, explain where the excess charge must reside on the conductor.

The previous activity suggests that the *excess charge on a conductor is located on the surface*. Let's think about this physically. Given the fact that like charges repel, the excess negative charges will repel each other, wanting to get as far away as possible from the other excess charges. Thus, it certainly seems plausible that the best way to do this is for the excess charges to spread themselves out over the surface of the conductor (see Fig. 20.8).

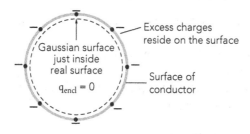

Fig. 20.8.

We just used Gauss's law to predict that excess charge on a conductor will reside on the surface. We now want to check this prediction, as well as think about what happens when the object is not a perfect, solid sphere. Suppose some excess negative charge is transferred to a metal can.[8] While the can is obviously not a solid sphere, we can think of it somewhat like a spherical shell with a hole on the top (see Fig. 20.9).

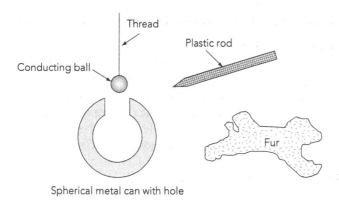

Fig. 20.9. Diagram showing the arrangement for Activity 20.6.2.

---

[8] Remember, since the can is a conductor and charges are free to move, it doesn't matter where you initially place the charges on the can. Also, although the diagram shows a spherically-shaped can, one can also use a cylindrically-shaped can.

20.6.2. Activity: Testing for Excess Charge on a Conductor

a. Assume that there is excess negative charge on a metal can that is either spherical or cylindrical in shape (with an opening at the top). In the following diagram, draw in where you think the excess negative charges will reside (this is just a *prediction*).

b. Now try the experiment. Use the fur to charge up the plastic rod, and then touch the rod to the metal can. Do this a few times to get plenty of charge on the can. Now lower the conducting ball into the center (the "inside") of the metal can as shown below. Go ahead and touch the ball to the bottom, inside surface of the can. What do you observe? Do you see any obvious electrical interaction?

c. There should be very little, if any, interaction between the ball and the inside of the can. This indicates that there is little, if any, excess charge residing on the "inside" surface of the can. Now try the *outside*. After charging the can again, bring up the conducting ball to the outside of the can and let it touch the outer surface, as shown below. What do you observe? Do you notice any obvious electrical interaction now?

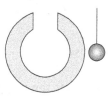

d. After touching the outside of the can, you should have seen the ball be strongly repelled from the can. This indicates that the ball was able to pick up some of the excess negative charge from the outer surface of the can, after which it was strongly repelled. For this can-shaped metal object, where does the excess charge appear to reside? Does this agree with your prediction? Based on your observations, sketch the location of the charges on the diagram below.

In Activity 20.6.1 we showed that any excess charge lies on the outer surface of a solid, spherical conductor. For a solid, spherical object, there is really only one surface (it's all an "outer surface"). For more complex objects, there may be both an "inner" and "outer" surface. In Activity 20.6.2 we saw that for a can-shaped, conducting object, (essentially) all of the excess charge lies on the outer surface of the can.

But what if we were to deform the can so that it resembled a plane more than a sphere? For a true plane, charge will reside on both sides (there is only an outer surface). For any object that has clear inner and outer surfaces (such as a solid sphere with a cavity inside), all the charge resides on the outer surface. And for an object that is in between, we can say that more of the charge will lie on the outer surface.

Charging by Induction

Using this same set-up, let's consider a slightly different experiment that demonstrates the concept of *electrostatic induction*. This type of experiment is generally referred to as the Faraday Ice Pail Experiment, based on an experiment Michael Faraday documented in 1843.

20.6.3. Activity: A Faraday Ice Pail Experiment

a. We will start with a thought experiment. Imagine you use the fur and rod to (negatively) *charge the conducting ball* but start with the *metal can uncharged*. The charged ball is now lowered into the center of the metal can *without* touching it. Do you think the can have a *net* charge on it? Will there be any charge on the inside of the can? Will there be any charge on the outside of the can? On the diagram below, predict the location of any charges that reside on the can.

You should have predicted that the negative charges on the ball repel the negative charges on the inside of the metal can. Because charges in the metal are free to move, the negative charges move toward the outer surface of the can, leaving the inner surface positively charged.[9] In other words, the conductor is *polarized*, similar to what we saw in Unit 19, but as long as the conducting ball has not touched the can, it will have no excess charge. If you were to touch the negatively-charged conducting ball to the metal can, negative charges would be transferred to the can, leaving it negatively charged as well (no surprises there). However, there is also a way to use this arrangement to leave the can with net *positive* charge.

---

[9] Even though we know it is the negatively-charged electrons that move, you can also think about the negatively-charged ball attracting positive charges in the can.

b. With the conducting ball hovering inside the can (but not touching it), is there anything you can do to leave the metal can with net positive charge? Talk this over with your partners to see if you can devise a method (it may not be obvious). **Hint**: You are free to touch the can.

c. Try the following experiment. While touching the outside of the can with your hand, slowly lower the charged ball until it is suspended inside the can (but not touching it). Now remove your hand from the can and then slowly extract the conducting ball, being careful not to let the ball touch the can. Check to see whether the can has been left with a positive net charge. Describe your procedure for checking and explain what you found.

You should have found that the can is left with excess positive charge; we say that the can has been charged by *induction*. To understand what's happening in this experiment, consider Fig. 20.10. Two *conductors* (A and B) begin uncharged but touching (left image). A negatively charged object is brought near the side of conductor A *without* touching it. The attractive/repulsive forces lead to a separation of charge (a polarization), and since objects A and B are touching the negative charges are pushed as far away as possible, moving onto object B and leaving net positive charge on object A (middle image). While the charged object is still nearby, objects A and B are separated. The charged object is now removed, but since A and B are no longer touching, they are left with equal but opposite charges (right image). Thus, we were able to create two charged objects without touching either one.

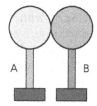 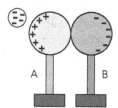 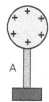 

Fig. 20.10. Diagram showing the process of charging by induction. Initially, two neutral conductors are touching (left). A third, charged object is brought nearby, leading to a separation of charges across the two conductors. The two conductors are then separated before removing the charged object, which leaves the conductors oppositely charged (the overall net charge is still zero).

GAUSS'S LAW CALCULATIONS

20.7   SPHERICALLY-SYMMETRIC, UNIFORM CHARGE DISTRIBUTION

In Activity 20.5.1 we went through the steps of using Gauss's law to calculate the electric field for the simple situation of a single point charge. We are now ready to use Gauss's law for more complicated charge distributions where we don't already know the answer. We begin with a spherically-symmetric, uniform charge distribution that consists of a solid sphere of radius R with a uniform charge density spread throughout its interior. Our goal is to compute the electric field at an arbitrary distance $r < R$ from the center, assuming a total excess charge of Q. Note that since $r < R$, the point in question is inside the sphere! But before digging in with Gauss's law, let's review the geometry for spheres and introduce the concept of charge density.

20.7.1.   Activity: Spherical-Shell Volume Element

a.   In Activity 20.5.1 we saw the surface area of a sphere of radius r is $4\pi r^2$. What is the equation for the *volume* of a sphere of radius r?

b.   Take the derivative of the volume V with respect to r. In other words, find dV/dr.

Note that dV/dr is equal to the surface area of a sphere! By rearranging this result, we can write $dV = 4\pi r^2 dr$. This is the volume element for a spherically-symmetric geometry and indicates how much the volume of a sphere would increase if the radius of the sphere were increased by some small amount dr. Figure 20.11 shows a cross-sectional view of a sphere that contains a small spherical shell in gray with radius r. The surface area of this shell is $4\pi r^2$, while its thickness is dr (dr is assumed to be so small that the difference in surface area between the inner and outer surfaces is negligible). The (approximate) volume of this shell is given by $dV = 4\pi r^2 dr$. These concepts will be useful when evaluating the integral in Gauss's law.

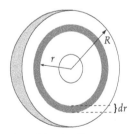

Fig. 20.11. Cross-section of a sphere showing a small, shaded volume element $dV = 4\pi r^2 dr$.

Charge Density

Gauss's law requires us to find the charge enclosed in our imaginary Gaussian surface. Since we now have a charge distribution containing many charges, the concept of charge density will help us determine the enclosed charge. For a three-dimensional object, the *charge density* is simply the charge per unit volume (note the similarity to mass density, which is the mass per unit volume). If the charge is uniformly distributed, the volume charge density ρ can be calculated as

$$\rho = \frac{Q}{V} \quad \left(\text{uniform volume charge density}\right)$$

where Q is the total charge and V is the total volume. In general, including cases where the charges are not uniformly distributed, the charge density can be a function of position and should be written in differential form as

$$\rho = \frac{dq}{dV}$$

where dq is the small amount of charge in volume dV.

20.7.2. Activity: Spherical Charge Density

a. Assume we have a solid sphere of radius R with a uniform charge density spread throughout its interior. Given that the charges lie throughout the volume, what type of material must the sphere be (insulator or conductor)? Why?

b. If the total charge contained in the sphere is Q, determine an expression for the charge density in terms of Q and R.

Now consider an imaginary Gaussian surface at some radius $r < R$ (see Fig. 20.12). To use Gauss's law, we need to determine the charge enclosed by this surface. We can use the differential form of ρ to calculate this by rewriting it as follows:

$$\rho = \frac{dq}{dV} \rightarrow dq = \rho\, dV$$

By integrating both sides from the center out to some arbitrary radius r we have

$$\int dq = q_{\text{encl}} = \int_0^r \rho\, dV$$

c. To calculate q_{encl}, we need to substitute ρ and dV into the expression above and integrate. Make those substitutions below (but wait to perform the integral until the next activity).

After that warmup, we are ready to proceed with using Gauss's law to calculate the electric field at an arbitrary distance $r < R$ from the center of a uniform, spherical charge distribution that has radius R and contains total excess charge Q (see Fig. 20.12). You will follow the same steps as in Activity 20.5.1, and we have summarized these for you in the activity below (you can also look back at Activity 20.5.1). Note that some of the steps may seem quite easy and straightforward—this is by design! The symmetry of the situation and the preliminary work we have done have us ready to move quickly through each step.

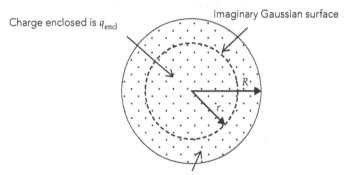

Fig. 20.12. Spherically-symmetric uniform charge distribution with a spherical-shell Gaussian surface.

20.7.3. Activity: Gauss's Law with Spherical Symmetry

We start with the equation for Gauss's law:

$$\oint \vec{E} \cdot d\vec{A} = \frac{q_{\text{encl}}}{\varepsilon_0}$$

Step 1 (done for you in Fig. 20.12!): Draw an (imaginary) Gaussian surface on the diagram with the *appropriate symmetry*. The Gaussian surface is a *spherical shell* of radius r located *inside* the charge distribution ($r < R$).

a. **Step 2**: Draw vectors representing both $\vec{E}$ and $d\vec{A}$ at different points on the imaginary Gaussian surface in Fig. 20.12. You must rely on the symmetry of the situation to determine the direction of the electric field. If you are unsure what the electric field looks like, be sure to ask!

b. **Step 3**: Determine the angle between vectors $\vec{E}$ and $d\vec{A}$ at all points on the imaginary Gaussian surface. Use this angle to evaluate the dot product. (Since there are not multiple faces, you don't need to worry about Step 3a.)

c. **Step 4**: Based on symmetry, should the magnitude of the electric field be constant over the entire surface? If so, use this fact to pull E out of the integral.

d. **Step 5**: Determine A, the surface area of the closed surface, and write down the full result for the integral on the left side of Gauss's law.

e. **Step 6**: Determine q_{encl}, the net charge enclosed by the Gaussian surface. Remember, this is the total charge that is *inside* (enclosed by) the imaginary surface. **Hint**: Use your result from Activity 20.7.2.

f. Step 7: Set the left side equal to the right side and solve for the quantity of interest (the electric field E). Finally, put in the direction of the field you determined in part (a) using the appropriate unit vector.

You should have found that the electric field is given by

$$\vec{E}(r) = \frac{kQr}{R^3}\hat{r} \quad \left(\text{for } r < R, \text{uniform spherical distribution}\right) \qquad (20.7)$$

where k is Coulomb's constant, Q is the total excess charge, R is the radius of the charge distribution, r is the distance from the center of the sphere, and $\hat{r}$ indicates that the field points radially outward for $Q > 0$ (the field would be flipped and point radially inward if $Q < 0$). We wrote the electric field as $\vec{E}(r)$ to make the dependence on r explicit. If you were provided values for the total charge and size of the distribution, you could calculate a numerical result for the electric field at any point (we'll keep everything in terms of variables).

20.7.4. Activity: Analyzing Our Result

a. What is the magnitude of the electric field exactly at the center of sphere? Does this make sense?

b. The magnitude of the electric field increases as r gets larger (as long as $r < R$). Explain why this seems physically reasonable for this situation.

c. What is the value of the electric field at $r = R$ (the outer edge of the distribution)?

Note that at the outer edge (the radius where $r = R$) the value of the electric field is exactly the same as if the entire charge distribution were a single point charge (of charge Q) centered at the origin (see Eq. (20.5)):

$$\vec{E}(r = R) = \frac{kQR}{R^3}\hat{r} = \frac{kQ}{R^2}\hat{r} = \vec{E}_p$$

In other words, if you only look at the field from *outside* the charge distribution, you can't tell the difference between this uniform, spherical distribution and the case when all the charge is concentrated in a point at the center. This same property holds true as you move further out ($r > R$) as well; once you are outside the distribution, the field is the same as for a point charge!

    d. In the space below, sketch a plot of the magnitude of the electric field as a function of radius r. Have the plot run from $r = 0$ out to some large $r \gg R$. **Note**: There will be a discontinuity in the slope of the graph at the point $r = R$ due to the charge distribution suddenly dropping to zero.

Hopefully this activity demonstrated the power of Gauss's law. We were able to determine the electric field at any point inside the charge distribution just by using symmetry and Gauss's law. Two facts bear repeating. First, were it not for the symmetry of the situation, we would not have been able to successfully apply Gauss's law without using a computer to do the calculations. Second, a uniform charge distribution like this only exists in an insulator (as we have seen, in a conductor all charges will be on the outer surface).

20.8 GAUSS'S LAW FOR OTHER SYMMETRIES

A Line of Charge

Both examples we looked at so far were for systems with spherical symmetry. What happens to the electric field when there is *cylindrical symmetry*? The system we will consider is shown in Fig. 20.13 and consists of a very long line of charge. We assume the line is essentially both infinitely long and infinitely thin (practically, this could be realized by using a very long, very thin thread). The point is that we can assume the line never ends, which allows us to not worry about what happens to the electric field near the ends of the line.

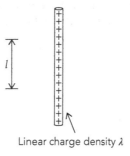

Linear charge density λ

Fig. 20.13. A line of charge with uniform charge density λ. While the line is assumed to go on forever, we will only consider the section of length l indicated.

Unlike our spherical charge distribution, which was finite in extent, we cannot assume a value for the total excess charge on the thread (it would be infinite, since the thread never ends!). Instead, we will assume we are given the *linear charge density*. The linear charge density is similar to the (volume) charge density ρ we saw earlier, but instead of a charge per unit volume, it is a charge per unit length. Linear charge density is typically represented using the variable λ and in SI units has units of coulombs per meter:

$$\lambda \equiv \frac{\text{charge}}{\text{unit length}} \quad \left(\text{linear charge density}\right)$$

In the next activity, you will use Gauss's law to find the electric field a perpendicular distance r from the line of charge. We will not rewrite the instructions for each step, but simply note the step and let you complete it (with hints provided). Feel free to refer back to earlier sections if you need reminders!

20.8.1. Activity: Gauss's Law with Cylindrical Symmetry

a. **Step 1:** Draw the Gaussian Surface. **Hints**: (1) Thinking about the symmetry, what shape of Gaussian surface would *symmetrically* enclose a line of charge? (2) Your Gaussian surface should only enclose the portion of the thread indicated by the length l. (3) Because your Gaussian surface must be closed, it will have multiple faces.

b. **Step 2:** Draw the Vectors $\vec{E}$ and $d\vec{A}$. **Hints**: (1) Think about the symmetry when determining which direction the electric field must point from the line of charge. (2) You need (at least) one $d\vec{A}$ on each face of the surface.

c. **Step 3:** Evaluate the Dot Product. **Hint**: You need to determine the angle between $\vec{E}$ and $d\vec{A}$ on each face.

d. **Step 4:** Move E out of the Integral (if appropriate). **Hint**: You only need to verify that E is constant on those face(s) where $\theta \neq 90°$, because the dot product is zero when $\theta = 90°$, and there is no contribution from that face.

e. **Step 5:** Evaluate the Integral. **Hints**: (1) Only include the area on those face(s) where $\theta \neq 90°$. (2) To find the area of the side, think about "un-rolling" the cylinder. (3) If you didn't already, you will want to label the radius and length of your cylindrical Gaussian surface (e.g., r and l).

f. **Step 6:** Determine the Charge Enclosed. **Hints**: (1) Because the linear charge density λ is assumed to be given, the symbol λ will be part of your answer. (2) Think about the units of λ and remember that we only want the charge *enclosed* by your Gaussian surface.

g. **Step 7:** Solve for the Electric Field. **Hints**: (1) Your result for the electric field should depend on the distance away from the line, but *not* the length of your Gaussian cylinder. (2) Don't forget to indicate the direction!

You should have found that

$$E(r) = \frac{\lambda}{2\pi\varepsilon_0}\frac{1}{r} = \frac{2k\lambda}{r} \quad \text{(uniform line of charge)} \qquad (20.8)$$

where the direction is radially away from the line of charge and in the second form we used the fact that $k = 1/4\pi\varepsilon_0$ (either form is perfectly valid). Note the electric field gets weaker as you move away from the line, but not as quickly as it did for a point charge ($1/r$ vs $1/r^2$). Also note that, unlike the uniform spherical distribution, the electric field goes to infinity as $r \to 0$. This is a limitation of our assumption that the line of charge is infinitely thin; any real line of charge would have some thickness, and our result only holds for values $r > r_{\text{line}}$.

A Plane of Charge

We have examined situations involving both spherical and cylindrical symmetry. There is one more geometry that can easily be solved using Gauss's law, and that is a plane of charge. We will consider a large plane (a sheet) with a uniform distribution of charge (see Fig. 20.14). Like with the line of charge, we will assume the plane is both infinitely large and infinitely thin so that we need not worry about what happens to the electric field near the edges. We will also assume we are given the (surface) charge density. *Surface charge density*, as

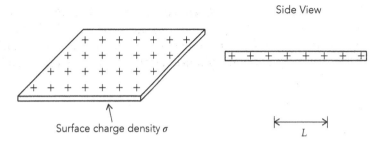

Side View

Fig. 20.14. A very large, very thin, uniformly-charged sheet. While the plane is assumed to go on forever, we will only consider the section of length L indicated.

you might expect, is a charge per unit area and is typically represented using the variable σ:

$$\sigma \equiv \frac{\text{charge}}{\text{unit area}} \quad \left(\text{surface charge density}\right)$$

In SI units, σ has units of coulombs per square meter. Our goal is to find the electric field some perpendicular distance r above (or below) the plane of charge.

20.8.2. Activity: Gauss's Law with Planar Symmetry

a. **Step 1**: Draw the Gaussian Surface. (There are a couple of possible choices. You may wish to ask your instructor if you have questions.)

b. **Step 2**: Draw the Vectors $\vec{E}$ and $d\vec{A}$.

c. **Step 3**: Evaluate the Dot Product on all surfaces.

d. **Step 4**: Move E out of the Integral (if appropriate).

e. **Step 5**: Evaluate the Integral.

f. **Step 6**: Determine the Charge Enclosed.

g. **Step 7**: Solve for the Electric Field. Be sure to specify the direction at the end.

You should have found that

$$\vec{E}(r) = \frac{\sigma}{2\varepsilon_0} \text{ "away from sheet"} \quad \text{(uniform plane of charge)} \quad (20.9)$$

Two important notes on this result. The first is that the direction is specified as being "away from the sheet" for $\sigma > 0$. Assuming the sheet is in the horizontal plane, the electric field points upward if you are located above the sheet and downward if you are located below the sheet. The second is that the electric field does *not* get weaker as you move away from the sheet; it remains constant!

This probably seems strange—how can the field not get weaker as you move farther and farther away? Doesn't this seem to violate some sort of energy principle? The reason, as is often the case, lies in one of our assumptions. In order to not worry about the electric field at the edges of the sheet, we assumed it was *infinitely* large. But this means that we have an infinite amount of charge contained on the entire sheet, which is not physically possible.

As you move farther from the sheet, the charges immediately below get farther away. Simultaneously, the electric fields from those charges off to the sides develop larger vertical components. If the sheet is infinitely large, this additional vertical contribution to the field makes up for the fact that you are moving farther away, and the electric field stays constant. Any physical situation will, of course, not be infinitely large, and our solution is only valid when the distance away from the sheet is small compared to the size of the sheet (when this is the case, the sheet "looks" big from the perspective of a point close to it).

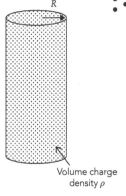

R

Volume charge
density ρ

20.9 PROBLEM SOLVING

A solid, insulating cylinder of radius R has a uniform volume charge density ρ distributed throughout. Your goal is to find the electric field (magnitude and direction) a distance r away from the central axis of the cylinder. You should assume the cylinder is very long so that you don't need to worry about the ends.

20.9.1. Activity: Field from a Uniformly-Charged Cylinder

a. Determine the electric field a distance r away from the central axis of the cylinder when $r < R$ (*inside* the cylinder).

b. Now determine the electric field a distance r away from the central axis of the cylinder when $r > R$ (*outside* the cylinder).

c. Sketch the magnitude of the electric field as a function of r from $r = 0$ out to some distance $r \gg R$.

Two views of a coaxial cable are shown in the figure below. The cable consists of a solid, cylindrical conductor in the center, surrounded by a cylindrical conducting shell on the outside, with a gap in between (typically filled with insulating plastic). If you have a cable modem for internet access, this is how data is transmitted to your network (this design is also used for other electrical connector cables).

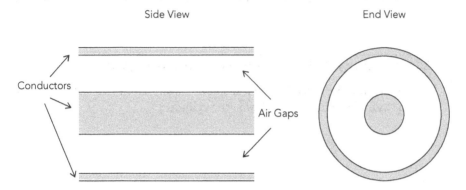

Assume there is a *linear* charge density of $+\lambda$ along the length of the inner conductor and $-\lambda$ along the outer conducting shell. Answer the questions below using the ideas we developed in this chapter, leaving your answer in terms of the problem variables and physical constants (when appropriate).

20.9.2. Activity: Coaxial Cable

a. Given that the inner, solid cylinder is a conductor, where must the excess charges on it reside?

b. The outer cylindrical shell is also a conductor and has negative excess charge. Given that there is excess positive charge on the inner cylinder, where do you predict the excess negative charge on the shell will reside? Briefly explain.

c. Determine the magnitude of the electric field in the region *within* the inner conductor. Briefly explain. **Hint**: This is quick!

d. Determine the magnitude of the electric field in the region *between* the conductors (in the gap). Give your answer as a function of the distance r from the central axis. **Hint**: Use our previous results.

e. Determine the magnitude of the electric field in the region *within* the outer conducting shell. Briefly explain. **Hint**: This is quick again!

f. Determine the magnitude of the electric field in the region *outside* the outer conducting shell. **Hint**: Think carefully about q_{encl}. Briefly explain.

UNIT 21: ELECTRIC POTENTIAL

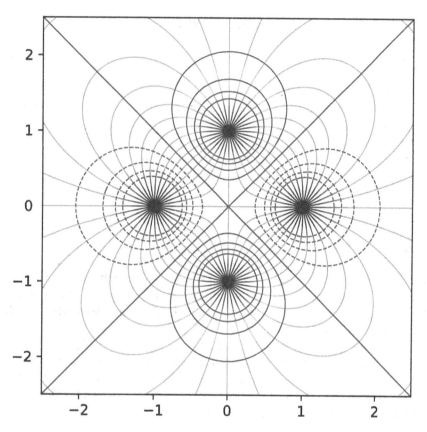

In the previous unit we defined the electric field and used Gauss's law to determine the field from distributions of electric charges. Although there are obvious differences between charge and mass, there is a similarity between the mathematical forms of the electrical forces between charges and the gravitational forces between masses. In analogy with gravitational potential energy, we can also define an electrical potential energy for charges. The figure above shows the electric field lines and equipotential surfaces for a four-charge configuration. Electrical potential energy is a key concept in understanding the behavior of the circuits that form an essential part of modern technology. In this unit we will study the concepts of electrical potential energy and electric potential, or voltage.

UNIT 21: ELECTRIC POTENTIAL

OBJECTIVES

1. To understand the mathematical similarities between gravitational and electrical forces.

2. To review the concepts of physical work and potential energy in a conservative field.

3. To understand the definition of electric potential, or voltage.

4. To learn how to determine electric field lines from equipotential surfaces and vice versa.

21.1   OVERVIEW

The enterprise of physics is ultimately concerned with mathematically describing the fundamental forces of nature. Nature offers us several of these forces, which include a strong force that holds the nuclei of atoms together, a weak force that governs certain kinds of radioactive decay in the nucleus, the force of gravity, and the electromagnetic force.

Two kinds of forces dominate our everyday reality—the gravitational force acting between masses and the Coulomb force acting between electric charges. The gravitational force allows us to explain how objects are attracted to Earth and how the Moon orbits Earth and the planets orbit the Sun. The breakthrough of Newton was to realize that objects as diverse as falling apples and orbiting planets are both moving under the action of the same gravitational force (see Fig 21.1).

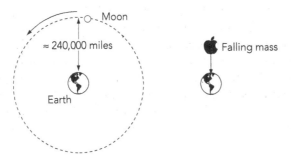

Fig. 21.1. The orbiting moon and a falling apple both experience a gravitational force that points toward the center of Earth.

As we saw in Units 6 and 19, the Coulomb and gravitational forces have the same mathematical form. Therefore, it is probably not surprising that, similar to the gravitational force, the Coulomb force allows us to describe how one charge "falls" toward another, or how an electron "orbits" a proton in a hydrogen atom (see Fig. 21.2).

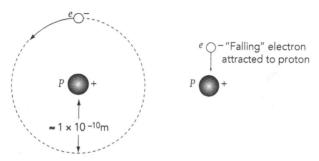

Fig. 21.2. An orbiting electron and a "falling" electron experience a Coulomb force that points toward the center of the proton.

We start this unit by exploring the mathematical symmetry between electrical and gravitational forces for two reasons. First, it can be beautiful to behold the unity that nature offers, allowing us to use essentially the same mathematics to predict the motion of planets and galaxies, the falling of objects, the flow of electrons in circuits, and the nature of the hydrogen atom and other chemical elements. Second, what we have already learned about the influence of the gravitational force on a mass, and the concept of potential energy in a gravitational field, can be applied to aid our understanding of the forces and energies for charged particles. In fact, a "gravitational Gauss's law" can be used to find the gravitational field in the presence of a large spherically symmetric mass.

We then introduce electrical potential energy, which is analogous to gravitational potential energy, as well as the concept of electric potential (commonly called voltage). A familiarity with these concepts will be essential for understanding the electric circuits and electronics we will study over the following units. Finally, we will examine equipotential surfaces for different configurations of charges.

ELECTRICAL AND GRAVITATIONAL FORCES

21.2   COMPARING ELECTRICAL AND GRAVITATIONAL FORCES

Let's start our discussion with a brief review of both the electrical and gravitational forces. The Coulomb force on charge q_B due to charge q_A is given in Fig. 21.3.

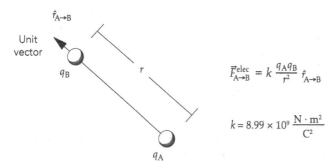

$$\vec{F}^{\text{elec}}_{\text{A}\rightarrow\text{B}} = k\,\frac{q_A q_B}{r^2}\,\hat{r}_{\text{A}\rightarrow\text{B}}$$

$$k = 8.99 \times 10^9\,\frac{\text{N} \cdot \text{m}^2}{\text{C}^2}$$

Fig. 21.3. The Coulomb force expressed in terms of the two charges, q_A and q_B, and the unit vector $\hat{r}_{A \rightarrow B}$.

Newton's discovery of the universal law of gravitation came after he first thought about planetary orbits. This was back in the seventeenth century, long before Coulomb began his studies. About the time of Coulomb's experiments with electrical charges, one of his contemporaries, Henry Cavendish, did a direct experiment to determine the gravitational force between two spherical masses in a laboratory. These experiments confirmed Newton's gravitational force law and allowed Cavendish to determine the gravitational constant G. As we saw in the final activity of Unit 6,[1] the law of universal gravitation describing the force on mass B due to mass A can be written as shown in Fig. 21.4.

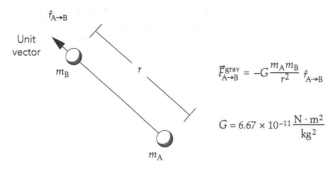

$$\vec{F}^{\text{grav}}_{\text{A}\rightarrow\text{B}} = -G\,\frac{m_A m_B}{r^2}\,\hat{r}_{\text{A}\rightarrow\text{B}}$$

$$G = 6.67 \times 10^{-11}\,\frac{\text{N} \cdot \text{m}^2}{\text{kg}^2}$$

Fig. 21.4. The gravitational force expressed in terms of the two masses, m_A and m_B, and the unit vector $\hat{r}_{A \rightarrow B}$.

---

[1] If you did not complete the optional activities at the end of Unit 6, you'll simply have to trust that this is the result!

21.2.1. Activity: The Electrical Versus the Gravitational Force

a. Examine the mathematical expressions for the two force laws given in Figs. 21.3 and 21.4. What is the same about the two force laws? What is different? For example, is the force between two like masses attractive or repulsive? How about two like charges? What part(s) of each equation determines whether the forces are attractive or repulsive?

b. We have seen that there are both positive and negative charges. What about masses—do you think negative mass could exist? If there were negative mass, how would a positive and negative mass interact with other?

Gravitational forces hold the planets in our solar system in orbit and account for the motions of matter in galaxies. Electrical forces serve to hold atoms and molecules together and govern how they interact. If we consider two of the most common subatomic particles, the electron and the proton, how do their electrical and gravitational forces compare with each other?

Let's peek into the hydrogen atom and compare the gravitational force between the two particles to the electrical force between them.[2] In order to do the calculation, you'll need to use the following constants.

$$Electron : m_e = 9.1 \times 10^{-31} \text{kg} \qquad q_e = -1.6 \times 10^{-19} \text{C}$$

$$Proton : \quad m_p = 1.7 \times 10^{-27} \text{kg} \qquad q_p = +1.6 \times 10^{-19} \text{C}$$

In addition, assume the distance between the electron and proton in a hydrogen atom is 1.0×10^{-10} m.

21.2.2. Activity: Electrical and Gravitational Forces in the Hydrogen Atom

a. Calculate the magnitude of the *electrical* force between the two particles. Is it an attractive or repulsive force?

----

[2] We will assume the classical (not quantum) picture of the hydrogen atom, with the electron and proton as two point particles some distance apart.

b. Calculate the magnitude of the *gravitational* force between the two particles. Is it an attractive or repulsive force?

c. Which is larger? By what factor? (Determine the ratio $F^{\text{elec}}/F^{\text{grav}}$.)

d. You should have found that for an electron and proton in a hydrogen atom, the electrical force is much, much stronger. But which force are *you* more aware of on a daily basis? Describe why you think this might be the case.

Because both force laws depend on distance in the same way ($1/r^2$), it is really the products $Gm_A m_B$ and $kq_A q_B$ that set the relative strengths of the two forces. For *microscopic* objects like the electron and proton, the masses tend to be very small, and the gravitational force can often be neglected. It's the opposite for *macroscopic* objects, where the (net) charges tend to be small (macroscopic objects are typically electrically neutral, comprised of nearly equal numbers of protons and electrons). Due to the incredibly large mass of Earth, it is usually the gravitational force we notice in the macroscopic world. That being said, there are plenty of situations where the net charges of macroscopic objects are not zero, at which point the strength of the electrical force becomes quite apparent (e.g., a balloon sticking to a wall due to electrostatic attraction)!

21.3 GRAVITATIONAL GAUSS'S LAW

From Unit 20, Gauss's law says the net electric flux through any closed surface is proportional to the net charge enclosed by the surface. Mathematically, this is represented by an integral of the dot product of the electric field with the area element vector over the closed surface:

$$\Phi^{\text{net elec}} = \oint \vec{E} \cdot d\vec{A} = \frac{q_{\text{encl}}}{\varepsilon_0} = 4\pi k q_{\text{encl}} \tag{21.1}$$

where $k = 1/4\pi\varepsilon_0$ is Coulomb's constant.

The similarity of the mathematical forms for electrical and gravitational forces suggests that there should also be a gravitational version of Gauss's law. For the electrical force $\vec{F}^{\text{elec}}$ we defined a corresponding electric field $\vec{E} = \vec{F}^{\text{elec}}/q_t$, where q_t is a "test charge" experiencing the force. As we saw in Unit 6, we can also define a gravitational field in terms of the gravitational force: $\vec{g} = \vec{F}^{\text{grav}}/m_t$, where m_t is a "test mass" experiencing the force.[3] In Activity 6.10.2 we found an expression for the gravitational field strength near the surface of Earth and generalized this to positions far from the Earth's surface. In this section we take a different approach and use a generalized Gauss's law to arrive at the same result. We start by determining the proper expression for the gravitational version of Gauss's law.

21.3.1. Activity: Gravitational Gauss's Law Equation

Using the form of Gauss's law containing Coulomb's constant k in Eq.(21.1), along with the forms of the two force laws, write down an analogous expression for a "gravitational Gauss's Law" using the appropriate gravitational quantities.

We can use this gravitational Gauss's law to calculate the gravitational field at some distance from the surface of Earth, just as we used the original Gauss's law to determine the electric field at some distance from a uniformly charged sphere. This is useful in figuring out the familiar force "due to gravity" near the surface of Earth, as well as at other locations not near the surface.

Figure 21.5 shows Earth surrounded by an imaginary Gaussian surface. We'll assume Earth is a perfect sphere with uniform mass density, and that the Gaussian surface is a spherical shell with radius $r > R_E$ (where R_E is the radius of Earth). We start with the equation for our new gravitational Gauss's Law determined in Activity 21.3.1:

$$\oint \vec{g} \cdot d\vec{A} = -4\pi G m_{\text{encl}} \tag{21.2}$$

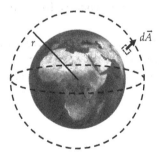

Fig. 21.5. Earth surrounded by a spherical Gaussian surface. We assume Earth is perfectly spherical and has uniform mass density.

---

[3] While $\vec{G}$ may seem a more natural choice to label the field, G (no vector) is already used for the gravitational constant, which has different units than the gravitational field. We choose $\vec{g}$ instead, although it does pose some potential confusion with "little g." As we will find, $g = 9.8$ N/kg is equal to $|\vec{g}|$ at the surface of the Earth.

This is a Gauss's Law problem with spherical symmetry, which is similar to situations we saw in Unit 20, so we won't go through every step in detail. Instead, we highlight a couple of the important points. As with the electrical case, by symmetry we expect both $\vec{g}$ and $d\vec{A}$ to point in the radial direction. However, given that the gravitational force is always attractive, we'll assume $\vec{g}$ points radially *in*ward, making $\theta = 180°$ everywhere in the dot product (giving us a negative sign).

21.3.2. Activity: The Gravitational Force of Earth

a. Looking at the left side of Eq. (21.2) why it is appropriate to pull $|\vec{g}|$ out of the integral?

b. On the right side, what is the total mass enclosed by our Gaussian surface? **Hint**: Remember that $r > R_E$.

Using your responses to parts (a) and (b), we can write down an expression for the gravitational field some distance r from the center of the Earth.

$$\vec{g} = -\frac{GM_E}{r^2}\hat{r} = -\frac{GM_E}{\left(R_E + h\right)^2}\hat{r} \qquad (21.3)$$

where M_E is the mass of Earth and the negative sign indicates that the gravitational field points inward toward the center of Earth (it's an *attractive* force). For the second form, we used the fact that $r = R_E + h$, where h is the height above the surface of Earth (the *altitude*). Notice that this result agrees with what we found in Activity 6.10.2!

c. Use your calculator to determine the strength of the gravitational field $|\vec{g}|$ at the *surface* of the Earth. Assume Earth has a radius $R_E = 6.38 \times 10^6$ m and a mass $M_E = 5.98 \times 10^{24}$ kg. Does the number look familiar? Briefly comment.

d. Imagine you used Eq. (21.3) to calculate $|\vec{g}|$ at the ceiling of the room ($h \approx 3$ m). How much would it differ from the value of $|\vec{g}|$ at the floor? Do you think you could measure the difference using the devices available in the classroom? **Hint**: You don't need to plug in numbers if you can describe what the result should show based on Eq. (21.3).

e. Finally, calculate $|\vec{g}|$ at the location of the Moon, which orbits approximately 384,000 km from the center of Earth.

21.4 WORK, ENERGY, AND CONSERVATIVE FORCES—A REVIEW

In preparation for tackling the idea of work and energy for situations involving charges and electric fields, let's review some definitions from mechanics. Work, like flux, is a *scalar* rather than a vector quantity (it does not involve a direction). From Eq. (10.8), work is defined mathematically as the dot product of a force and a displacement integrated over a path from some initial location $\vec{r}_1$ (point 1) to some final location $\vec{r}_2$ (point 2):

$$W^{1 \to 2} = \int_{\vec{r}_1}^{\vec{r}_2} \vec{F} \cdot d\vec{r} \tag{21.4}$$

Remember that the dot product can be written as $\vec{F} \cdot d\vec{r} = F \, dr \cos \theta$, where θ is the angle between $\vec{F}$ and $d\vec{r}$. In principle, both $\vec{F}$ and θ can change at every location along the chosen path (see Fig. 21.6).

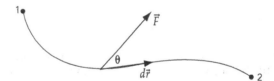

Fig. 21.6. A curved path from point 1 to point 2. At any point along the path a small test mass can experience a changing force $\vec{F}$ that's in a different direction than the small displacement $d\vec{r}$.

Let's review the procedure for calculating work. For any given force and displacement, one can calculate the work done by that force. For example, Fig. 21.7 shows a cart of mass m moving from point a at the bottom of a ramp to point c at the top of the ramp. Imagine you move the cart up the ramp by grabbing it and pulling at a constant velocity in the direction indicated by the arrow. The cart moves a total distance L (the length of the ramp) while increasing its height by an amount h. *We will treat the cart as a point mass and ignore any friction.*

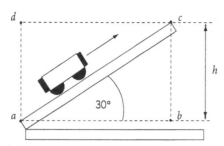

Fig. 21.7. A mass moving up a 30° incline from point *a* to point *c*. The total length of the ramp is *L*, while the difference in height between points *a* and *c* is *h*.

21.4.1. Activity: Mechanical Work—A Thought Experiment

a. How many forces are acting on the cart in Fig. 21.7 while you pull it up the ramp? List the forces and draw the corresponding force vectors in Fig. 21.7. Do any of these forces change during the motion, or do they remain constant? Remember that we are treating the cart like a point mass, ignoring friction, and pulling the cart at a constant velocity.

b. Relate the length of the ramp L to the height of the ramp h using geometry/trigonometry.

c. We can calculate the work done by each force during the motion. Let's start with the normal force—calculate how much work is done by the normal force (W_N) as you pull the cart up the ramp. Remember, you are pulling directly up the ramp at a constant velocity.

d. Next, calculate the work done by your applied force (W_{app}) during the motion. Leave your answer in terms of the problem variables. **Hint**: You will need to determine the magnitude of the applied force required to keep the cart moving at a constant velocity.

e. Calculate the work done by the gravitational force (W_{grav}) during the motion. Be careful with signs.

f. Finally, calculate the *net work* done during the motion.

You should have found that the net work done was zero: the positive work you did with the applied force was exactly canceled by the negative work done by the gravitational force (the normal force does no work). This is consistent with the *work-energy principle* for a point mass from Unit 10:

$$W_{\text{net}}^{1 \to 2} = \int_{\vec{r}_1}^{\vec{r}_2} \vec{F}_{\text{net}} \cdot d\vec{r} = \Delta K \tag{21.5}$$

where ΔK is the change in kinetic energy of the mass. Because the cart moves at a constant velocity the entire time, we know that $\Delta K = 0$.

Potential Energy

Previously, we determined that gravity is a *conservative force*, meaning that the work done by gravity when going from one point to another is independent of the path taken. Thus, for the cart in Fig. 21.7, we could just as well have lifted the cart straight up from point $a \to d$, and then moved horizontally from point $d \to c$; the work done would be the same as along the direct path from $a \to c$. Moreover, because the work done is path independent, we found it convenient to define the change in gravitational potential energy as

$$\Delta U_g \equiv -W_g \tag{21.6}$$

For the gravitational force near the surface of Earth, the change in potential energy is related to the change in vertical position (height) of the mass:

$$\Delta U_g = +mg\Delta y \quad \left(\text{near surface of Earth}\right) \tag{21.7}$$

where m is the mass, $g = 9.8$ N/kg is the gravitational field strength near the surface of Earth, and $\Delta y = y_{\text{final}} - y_{\text{initial}}$ is the change in vertical position. The following activity provides a brief review of mechanical energy.

21.4.2. Activity: Mechanical Energy Review

a. Use your result from part (e) of Activity 21.4.1 and Eq. (21.6) to calculate the change in gravitational potential energy of the cart as it is moved from point a to point c along the ramp. Does this agree with what you expect based on Eq. (21.7)?

b. Now imagine moving the cart *horizontally* from point $d \rightarrow c$. Based on the expression $\Delta U_g = mg\Delta y$, the work done by gravity along this path should be zero. Explain why this is to be expected based on the equation for work: $W_g^{d \rightarrow c} = \int_d^c \vec{F}_g \cdot d\vec{r}$.

c. Finally, consider again the full motion up the ramp from $a \rightarrow c$. Calculate the total change in mechanical energy ΔE_{mech} during this motion.

21.5 WORK AND THE ELECTRIC FIELD

We just reviewed some of the basic ideas about work and energy for a mechanical system, focusing on the conservative force of gravity. Given the mathematical similarity between the Coulomb force and the gravitational force, it should come as no surprise that experiments confirm that the force on a charge from an electric field is also conservative. Thus, the work needed to move a charge from point 1 to point 2 is independent of the path taken, so we can define an (electrical) potential energy associated with the electric force. Based on analogy with the gravitational force, we can easily write down the equation for the work done by the electrical force on a small test charge q_t traveling between points 1 and 2. We just need to substitute in the appropriation force $\vec{F}_{\text{elec}} = q_t\vec{E}$ into the equation for work:

$$W_{\text{elec}}^{1 \rightarrow 2} = \int_{\vec{r}_1}^{\vec{r}_2} \vec{F}_{\text{elec}} \cdot d\vec{r} = \int_{\vec{r}_1}^{\vec{r}_2} q_t\vec{E} \cdot d\vec{r} \qquad (21.8)$$

Determining the work done by gravity near the surface of Earth is straightforward, as the gravitational force is assumed to be perfectly uniform in this region. In other words, near the surface of Earth, the gravitational field has constant strength $\left(g = |\vec{g}| = 9.8 \text{ N/kg}\right)$ and always points in the same direction ("down" toward the center of Earth). As we begin our discussion of work done by electrical forces, we will assume a similarly uniform electric field. In this case, both q_t and $|\vec{E}| = E$ can be pulled out of the integral in Eq. (21.8).

21.5.1. Activity: Work Done on a Charge in a Uniform Electric Field

a. Assume a test charge q_t travels a distance d between points 1 and 2 such that the path is *parallel* (or antiparallel) to a uniform electric field of magnitude E (see Fig. 21.8). Calculate the work done by the electric force for each of four cases below. Be careful with signs.

(i) q_t is positive and moving from $1 \rightarrow 2$.

(ii) q_t is positive and moving from $2 \rightarrow 1$.

(iii) q_t is negative and moving from $1 \rightarrow 2$.

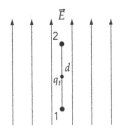

Fig. 21.8.

(iv) q_t is negative and moving from $2 \rightarrow 1$.

b. Equation (21.6), $\Delta U_g \equiv - W_g$, defines the change in gravitational potential energy in terms of the work done. A similar relationship is also valid for the case of an electrical potential energy: $\Delta U_{elec} \equiv - W_{elec}$. Calculate the change in electrical potential energy for each of the four cases described in part (a).

Note the similarity of your results in parts (a) and (b) to that of a test mass m_t moving a vertical distance d near the surface of Earth. The magnitudes of the expressions should have the same form: $q_t E d$ versus $m_t g d$. But one needs to be especially careful in the electric case because, unlike with mass, the charge can be both positive and negative!

c. Now assume the small test charge q_t travels from point 3 to point 4 in the same uniform electric field, but this time the path is *perpendicular* to the field lines (see Fig. 21.9). What is the work done by the electric force during this motion? Does it matter whether the charge is positive or negative? What about if you move in the opposite direction (from 4 to 3)?

As in the gravitational situation, the work (and change in potential energy) depends on the dot product of the force and displacement, so the angle between the force and displacement is important when calculating the work. The situation depicted in Fig. 21.9 is akin to moving a mass horizontally near the surface of Earth. Unless the height of the mass changes, no work is done, and the gravitational potential energy remains constant. Visually, one can picture gravitational field lines (or vectors) that point vertically down; if the path moved is perpendicular to these gravitational field lines ($\theta = 90°$ in the dot product), no work is done. Similarly, if a charge moves perpendicular to the electric field lines, no work is done, and the electrical potential energy remains constant.

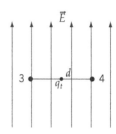

Fig. 21.9.

ELECTRIC POTENTIAL DIFFERENCE

21.6 ELECTRICAL POTENTIAL ENERGY VERSUS ELECTRIC POTENTIAL

By combining Eqs. (21.6) (for the electric case) and (21.8), we can write out an expression for the change in electrical potential energy:

$$\Delta U_{\text{elec}}^{1 \to 2} = -W_{\text{elec}}^{1 \to 2} = -\int_{\vec{r}_1}^{\vec{r}_2} \vec{F}_{\text{elec}} \cdot d\vec{r} = -\int_{\vec{r}_1}^{\vec{r}_2} q_t \vec{E} \cdot d\vec{r} \qquad (21.9)$$

Recall that we originally defined the electric field as a force per unit charge: $\vec{E} = \vec{F}_{\text{elec}}/q_t$, where q_t is a small test charge and $\vec{F}_{\text{elec}}$ the (electric) force on that charge. The electric field is a useful concept, as it allows us to think about the effect of the electric interaction independent of any particular test charge. Motivated by this, we can also define a new quantity that is the *change in electrical potential energy per unit charge*, which we represent with the variable V:

$$\Delta V^{1 \to 2} \equiv \frac{\Delta U_{\text{elec}}^{1 \to 2}}{q_t} = -\int_{\vec{r}_1}^{\vec{r}_2} \vec{E} \cdot d\vec{r} \qquad (21.10)$$

Note that $\Delta V^{1 \to 2}$ only depends on the movement through the electric field, not the charge that is being moved. The quantity V is called the *electric potential*, and so $\Delta V^{1 \to 2}$ is the *electric potential difference* between points 1 and 2. Admittedly, this is a confusing choice of term, as it is very similar to U, the electrical potential *energy*. Fortunately, $\Delta V^{1 \to 2}$ is often referred to by its more common name *voltage*, which comes from the units of electric potential difference: $\Delta V^{1 \to 2}$ has units of joules per coulomb, which is defined to be a *volt* (1 V $\equiv$ 1 J/C).

If the electric field is *uniform* (same magnitude and direction everywhere) over a region, Eq. (21.10) can be simplified by moving $\vec{E}$ out of the integral.

In particular, we consider the voltage change between points 1 and 2 for the case of a uniform electric field and straight path:

$$\Delta V^{1\to2} = -\vec{E} \cdot \int_{\vec{r}_1}^{\vec{r}_2} d\vec{r} = -E\,|\Delta\vec{r}|\cos\theta \quad \text{(uniform field, straight path)}$$

where $|\Delta\vec{r}| = |\vec{r}_2 - \vec{r}_1| \equiv d$ is the distance between points 1 and 2 and θ is the angle between $\vec{E}$ and $\Delta\vec{r}$. In the situation where $\vec{E}$ and $\Delta\vec{r}$ point in the same (or opposite) direction, $\cos\theta = \pm1$, and this simplifies to $\Delta V = \pm Ed$, where the sign is negative when moving in the direction of $\vec{E}$ and positive when moving opposite the direction of $\vec{E}$.

Potential Energy for Point Charges

We begin by considering the situation of two point charges, one of which is called q_A and is fixed at the origin of our coordinate system. We then imagine taking a second test charge q_t, starting it very far away from q_A, and then slowly bringing it in toward q_A (see Fig. 21.10).

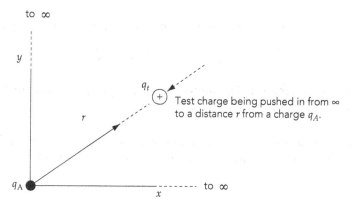

Fig. 21.10. A (positive) point charge located at the origin and a (positive) test charge being brought in from infinitely far away.

When q_t is really far away, there is essentially no interaction between the two charges. But as we move q_t closer to q_A, it will experience an increasing Coulomb force due to the presence of q_A.[4] We can use Eq. (21.9) to calculate the change in electrical potential energy that occurs when we move q_t in toward q_A, starting at point 1 (far away) and reaching an arbitrary point 2 that is closer to q_A. For simplicity, we assume both charges are positive and lie along the x-axis, giving

$$\Delta U_{\text{elec}}^{1\to2} = -\int_{\vec{r}_1}^{\vec{r}_2} \vec{F}_{\text{elec}} \cdot d\vec{r} = -\int_{\vec{r}_1}^{\vec{r}_2} \frac{kq_A q_t}{x^2}\hat{x} \cdot d\vec{r}$$

where we have plugged in for $\vec{F}_{\text{elec}}$ using Coulomb's law and noted that since both charges lie on the x-axis, $\hat{r}_{A\to t}$ in Coulomb's law points in the $+\hat{x}$ direction. In the next activity, you will perform the integral to determine the electrical potential energy associated with this situation.

---

[4] By Newton's third law, of course, q_A feels an equal and opposite force. But for this activity we are assuming q_A is fixed at the origin by some external force and focus our attention on the test charge q_t.

21.6.1. Activity: Potential Energy of Two Point Charges

a. The integral contains the dot product $\hat{x} \cdot d\vec{r}$. Remembering that we can write $d\vec{r}$ in Cartesian coordinates as $d\vec{r} = dx\,\hat{x} + dy\,\hat{y} + dz\,\hat{z}$, determine the result of this dot product.

b. Pull out all the constants and perform the integral, being sure to evaluate the result at the end points. **Hint**: If you don't remember how to do this integral, note that $1/x^2$ can be written as x^{-2}, which then looks like a "power law" integral.

After evaluating the result at the end points, you should have found that

$$\Delta U_{\text{elec}}^{1 \to 2} = \frac{kq_A q_t}{x_2} - \frac{kq_A q_t}{x_1}$$

where x_1 is the initial location of q_t (far away) and x_2 is the final location of q_t (closer to q_A).

c. We are assuming x_1 is very far out on the positive x-axis. What happens to the second term in our result if we assume $x_1 \to +\infty$?

Before writing the final form of our result, we make three additional notational changes for simplicity (and to follow convention). First, the final location of the test charge x_2 is arbitrary; we are moving q_t closer and closer to q_A and are free to stop moving at any point. Therefore, we might just as well replace the specific value of x_2 with the general variable x.

Next, your answer to part (c) should make it clear that the second term in our expression for $\Delta U_{\text{elec}}^{1 \to 2}$ will likely have a small magnitude compared to the first term. In fact, remember that our calculation only gives a *change* in potential energy (ΔU) due to the movement of q_t. But as with gravitational potential energy, we are free to set the location where the potential energy is zero.[5] For our situation of point charges, the convention is to choose the potential energy to be zero when the test charge is infinitely far away (effectively dropping the

---

[5] Remember Activity 11.3.2, where the College President was dropping water balloons out the building window. You (on the ground) and the President (up on a higher floor) could independently choose where the gravitational potential energy was equal to zero. In other words, the potential energy is only defined up to an arbitrary constant.

second term). This choice leads to $U_{elec} = kq_Aq_t/x$, where we have also dropped the "change in U_{elec}" notation (along with the $1 \rightarrow 2$ superscript) since we have assumed that point 1 is infinitely far away and point 2 is at some arbitrary position x.[6]

Finally, because there is nothing special about the x-direction (the charges could just have easily been along the y-axis, or along some diagonal), it is customary to write the denominator as r, which represents the *separation distance between the two charges*:

$$U_{elec} = \frac{kq_Aq_t}{r} \quad \text{(potential energy of two point charges)} \quad (21.11)$$

Although we assumed that q_t was a test charge and that both charges were positive, this expression is valid for any two charges q_A and q_B (positive or negative, and with any magnitude).

21.6.2. Activity: Analyzing Our Result

Compare Eq. (21.11) to the expression for the *electric force* between two charges (Coulomb's law), assuming the two point charges are labeled q_A and q_t. What are the similarities and differences between the two equations?

Potential Energy for a Configuration of Charges

If there are *more* than two point charges, the total potential energy of the configuration is determined by adding up the potential energies from *each pair* of charges. For example, with three point charges (labeled q_A, q_B, and q_C), the potential energy of the three-charge system would be

$$U_{elec}^{total} = \frac{kq_Aq_B}{r_{AB}} + \frac{kq_Aq_C}{r_{AC}} + \frac{kq_Bq_C}{r_{BC}}$$

where r_{ij} is the separation distance between charges q_i and q_j. Note that since U_{elec} is not a vector, we can just add these numbers together without worrying about direction!

Electric Potential for a Point Charge

We found the electrical potential energy of a two-charge configuration and noted it has similarities with the formula for the electric force between two charges. But we have seen that it is often more convenient to discuss the "force per unit charge," or the electric field of a single charge instead of the electric force between two charges. Similarly, it will be useful going forward to consider the

---

[6] It is analogous to say that the gravitational potential energy of a mass at some height h near the surface of Earth is simply mgh. Implied in this statement is the fact that you are actually referencing the potential energy of the mass-Earth system and that your reference point (or ground) is defined at $h = 0$.

"(electrical) potential energy per unit charge," or the electric potential of a single charge, as defined in Eq. (21.10). We derive this formula in the following activity.

21.6.3. Activity: The Electric Potential of a Point Charge

a. Use Eq. (21.10) (left part only) and Eq. (21.11) to determine "the potential" of a single charge q_A. For notational consistency, you will want to drop the "change in" notation (along with the $1 \to 2$ superscript) in Eq. (21.10), just like we did for the potential energy in Eq. (21.11). **Hint**: This is straightforward, so no need to overanalyze.

b. Compare your result to that for the *electric field* of a single point charge q_A. What are the similarities and differences between the two equations?

21.7 POTENTIAL DUE TO A CONTINUOUS CHARGE DISTRIBUTION

The electric potential V, just like the electrical potential energy U, is a scalar quantity (i.e., *not* a vector). This is one of the main reasons for defining such a quantity; when combining the (electric) potential from multiple point charges you can simply add up the numbers without worrying about directions or vector components. As in Unit 19 for the electric field, the electric potential from a continuous charge distribution can be calculated several ways, and each method should yield approximately the same result. For example, we could sum up the contributions from several finite charge elements, or we could use an integral method in which the potential dV from each element of charge dq is integrated to give a total potential at the location of interest.

Let's consider the same, simple charge distribution as in Section 19.7: a cylindrical rod that has a uniform charge distribution throughout (see Fig. 21.11). We are interested in finding the (electric) potential V at point P located a distance d from the end of the rod. The rod has a total charge Q and length L, and we have oriented our coordinate system so that both point P and the rod are along the x-axis.

We let x represent the distance between a point on the rod and point P, and this distance changes as you move along the rod. If we consider the rod to be a series of point charges, we would need to sum up the potential at point P due to each charge using our result from Activity 21.6.3. Instead, we take the integral

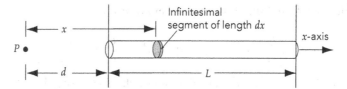

Fig. 21.11. An infinitesimal length dx at an arbitrary distance along a charged rod of length L. The distance from point P to the end of the rod is d, while the distance to the given infinitesimal segment is denoted as x.

approach (similar to Activity 19.7.3), so this sum becomes an integral with the appropriate substitutions:

$$V^{tot} = \sum_i \frac{k\Delta q_i}{x_i} \rightarrow \int_{x_{min}}^{x_{max}} \frac{k\,dq}{x}$$

where dq is the infinitesimal amount of charge in each segment, and the integral is performed over the length of the rod. Note that because V is not a vector, there are no unit vectors to worry about!

21.7.1. Activity: Potential from a Charged Rod

a. It is x that is changing as we integrate along the rod, so it would make sense that there should be a dx inside the integral. There is, but it is once again hidden in the dq. Use the total charge Q and the rod length L to determine how the amount of charge dq depends on dx. Keep everything in terms of the problem variables throughout this activity.

b. Based on the diagram, determine the values of x_{min} and x_{max}. Remember, the integral should go from one end of the rod to the other.

c. Substitute your expressions for dq, x_{min}, and x_{max} into the integral and perform the integration, including evaluating it at the end points. What is the final expression for the magnitude of the total electric potential at point P?

You should have found that the potential at point P is

$$V_{\text{rod}}^{\text{tot}} = \frac{kQ}{L}\left[\ln\left(d+L\right) - \ln\left(d\right)\right] = \frac{kQ}{L}\ln\left(\frac{d+L}{d}\right)$$

where point P is a distance d from the end of the rod. Note that this distance d is arbitrary (point P could be anywhere), and so one could consider d to be a variable that can change.[7] In other words, V can be considered a function of d: $V(d)$. It turns out to be of interest to determine how the potential changes as d is varied. Mathematically, this involves taking the derivative of $V(d)$ with respect to the variable d. In the following activity, we will compute this derivative and see where it leads.

21.7.2. Activity: The Derivative of the Potential

a. Take a derivative of your result from Activity 21.7.1 with respect to the variable d (be careful, this is notationally very awkward).

b. Compare your result to what we found for the magnitude of the electric field in Activity 19.7.3: $E^{\text{tot}} = kQ/d(L+d)$. What do you notice? You may need to simplify your result from part (a), and factoring out a negative sign might help you see the relationship better.

Getting the Field from the Potential

You should have found that the derivative with respect to distance d of the potential gives the same result as we found in Activity 19.7.3 (apart from an overall negative sign). It turns out that for a one-dimensional situation one can obtain the electric field directly from the potential by taking the derivative and adding a negative sign! Examining Eq. (21.10) this is probably not surprising, since there we found that the potential was an integral of the electric field (with an extra minus sign).

In three dimensions the operation to go from the potential (a scalar) to the electric field (a vector) is called the *gradient* and is denoted as

$$\vec{E} = -\vec{\nabla}V$$

where $\vec{\nabla}$ is known as the "del" operator. This topic is typically reserved for upper-level physics courses, so we will not go into any details beyond noting that one can obtain the electric field from the potential by taking the derivative, adding a minus sign, and putting in the appropriate direction.

---

[7] Alternatively, one could substitute in x for d in the final result and consider how the potential changes with x. This is the more traditional approach mathematically, but we avoid doing so here since it involves a redefinition of the variable x (we already defined it to be the distance from point P to any position along the rod).

21.8 EQUIPOTENTIAL SURFACES

When considering the gravitational potential energy, we saw that it is possible to move a mass without doing any physical work. Thus, one remains at the same gravitational potential energy anywhere along such a path. The same thing is true in the electrical case; it is possible to move an electric charge along a path without doing any work. Along such a path the electrical potential remains the same, and the path is called an *equipotential path* (or *equipotential surface* more generally). In this section we consider equipotential surfaces and their relationship to electric field lines. To perform the experiment in Activity 21.8.3, you will need the following equipment:

- 2 carbonized-paper equipotential plotting sheets with different conductor configurations (or a water tray with different conductors and graphing sheets)
- 1 voltmeter with leads
- 1 battery (6 V typically works well)
- 2 alligator clip leads

21.8.1. Activity: Electric Field Lines and Equipotentials, Part 1

a. For the *gravitational* case, how can you move a test mass such that the gravitational force does no work (and therefore the gravitational potential energy remains constant)? Explain how you know.

b. Hopefully, you said horizontally, or something equivalent. From Eq. (21.4), the work depends on the dot product of the force and the displacement, and the gravitational force points straight down (toward the ground). In particular, what is the angle θ between the gravitational force vector and the displacement in the situation where the gravitational force does no work?

As we've discussed, one can consider the gravitational force arising from a gravitational field $\vec{g}$, which also points straight down near the surface of Earth. When a test mass is moved perpendicular to the gravitational field lines, no work is done, and the gravitational potential remains constant. Now let's consider the analogous electrical case, which can be created by having two uniformly-charged plates held a fixed distance apart. The side view of this situation is shown in Fig. 21.12. Note that the electric field is uniform and pointing straight "down," just like the gravitational field near the surface of Earth.

c. How could you move a test charge so that the electric force does no work? Draw in this path on the diagram below. In three-dimensions, what are the actual shapes over which you could move the charge with no work being done?

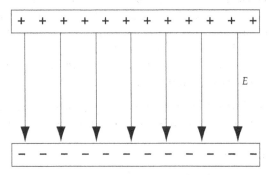

Fig. 21.12. Diagram showing a uniform electric field pointing straight down. The field is created by two uniformly-charged plates, and the image shows a side view.

For a uniform field as in Activity 21.8.1, the equipotential surfaces are *planes that are perpendicular to the electrical field* (i.e., horizontal planes in Fig. 21.12). A test charge can move anywhere in a horizontal plane and the electric potential will remain constant. Why? Because no matter how it moves in this plane, the angle between its displacement and the electric field is 90°, and so $\vec{E} \cdot d\vec{r} = 0$. However, if the charge moves out of this plane, some component of its displacement will be along the direction of the electric field (or opposite it), work will be done, and the potential will change.

Note that there are infinitely many of these equipotential surfaces—you can start the test charge at any "height" between the two charged plates. For any starting height, there is a horizontal plane that serves as the equipotential surface for that test charge. Every one of the infinitely-many equipotential surfaces is at a slightly different potential. And since $\Delta U_{elec}^{1\to2} = q_t \Delta V_{elec}^{1\to2}$, we can think of these surfaces as labeling the potential energy of the test charge; movement on a surface keeps the potential energy the same, while movement between surfaces implies a change in potential energy.

21.8.2. Activity: Electric Field Lines and Equipotentials, Part 2

a. Consider the situation depicted below of a point charge with its electric field lines. What path could you move a second test charge without doing any work? Therefore, what is the shape of an equipotential surface around a point charge (in two dimensions on the page)? Draw in some of these surfaces. In three dimensions, what are the shapes of the equipotential surfaces around a point charge?

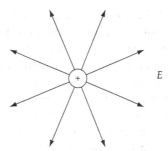

b. Now, consider the diagram below that shows a positive and negative charge in an "electric dipole" configuration. The electric field lines for this configuration are shown. Draw in some equipotential surfaces on the diagram. **Hint**: This might be a little more challenging—remember that you can always rely on needing to keep $\vec{E} \cdot d\vec{r} = 0$ as the test charge moves!

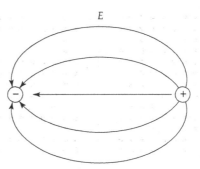

Measuring Equipotential Surfaces

Finally, we wish to experimentally explore equipotential surfaces in the region around different charge configurations. In particular, we will be able to test one or more of your predictions from Activities 21.8.1 and 21.8.2. It is typically easiest to examine and map two-dimensional versions of the charge configurations, meaning that the equipotential "surfaces" will actually be one-dimensional lines. There are different methods for performing this test, including using carbonized paper, a water tray, or a computer simulation.[8] In the discussion that follows, we describe an experiment using carbonized paper, but your instructor may have you use a different approach (the steps are easily generalized to the other methods).

The basic idea is to map out the equipotential surfaces given some configuration of charges, which means we will need both a source of charges and a way to measure the electric potential. We can wire up a battery to put charges onto one or more conductors and then use a *voltmeter* to measure the electric

---

[8] For example, the PhET site at the University of Colorado has a simulation that allows one to map out equipotential surfaces.

potential at different points in space. By tracing out paths of constant potential, we should be able to map out the equipotential lines for the configuration.

For the carbon-paper experiment, we will use pieces of carbonized paper with conducting metal painted on them in different shapes to *simulate* a charge configuration. For example, Fig. 21.13 shows two different patterns: the left paper has two small circular conductors painted on it, while the right paper has two linear conductors. Wiring up the battery to the two conductors will create charge configurations analogous to the dipole of Activity 21.8.2.b (for two circular conductors) or the uniform field of Activity 21.8.1 (for two linear conductors). Your measurement of the equipotential surfaces should convince you that the field takes the shape you expect.

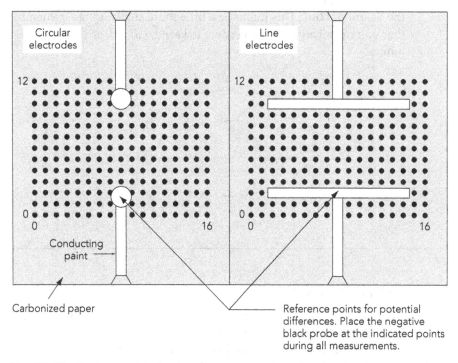

Fig. 21.13. Equipotential plotting sheets consisting of carbonized papers with point and line electrodes painted on them with conducting paint.

Choose a configuration of conductors. Use the alligator clips to connect the terminals of the battery to each of the electrodes as shown in Fig. 21.14. Turn on the voltmeter.[9] Set the tip of the black voltmeter probe (plugged into the COM input) on the center of the negative electrode. Place the red probe (plugged into the V-Ω input) on any location on the paper. The reading on the voltmeter will show you the electric potential difference between the two points ($\Delta V^{1 \to 2}$).

---

[9] We will use the digital voltmeter more extensively in Unit 22, and so we save a full description of it until then.

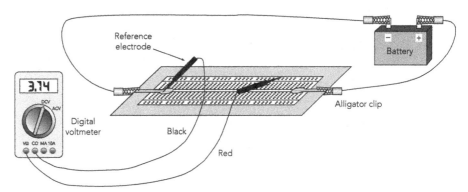

Fig. 21.14. Suggested set-up for measuring potential differences between various points on equipotential plotting sheets.

21.8.3. Activity: Mapping the Equipotentials

a. Place the black probe on the center of the negative electrode and then use the red probe to find a spot on the carbonized paper where there is a potential difference of approximately 1.0 V. Now, move the red probe in such a way that the potential difference stays (roughly) constant at 1.0 V. What path does the red probe trace out as it moves if the potential stays constant? Be sure to note which configuration of electrodes you are using.

b. Do something similar for a potential difference of 2.0 V, 3.0 V, etc. Make a sketch of the equipotential lines for the different potential differences you mapped (either directly on Fig. 21.13 or in the space below).

c. Compare your results to those you predicted in Activity 21.8.1 and/or 21.8.2. Is it what you expected? If not, explain how and why it differs. **Note**: Be sure to turn the voltmeter *off* when you are finished with your observations.

21.9 PROBLEM SOLVING

Living cells "pump" singly-ionized sodium ions Na⁺ from the inside of the cell to the outside to maintain an electric potential difference across the cell membrane that has a magnitude of approximately 100 mV:

$$\Delta V_{\text{membrane}} = V_{\text{out}} - V_{\text{in}} = +100 \times 10^{-3} \text{ V}$$

The term pump is used because work must be done to move a positive ion from the negatively-charged inside of the cell to the positively-charged outside. This must go on continuously because sodium ions "leak" back through the cell wall by diffusion.

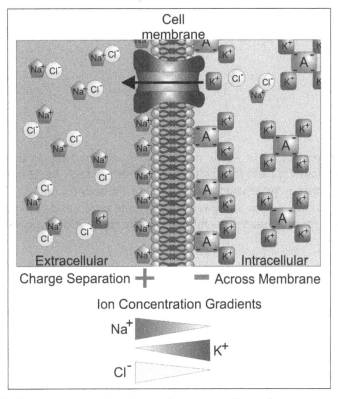

Fig. 21.15. Electric potential maintained across a cell membrane. Positively-charged sodium ions line the outside of the cell membrane, while negatively-charged ions line the inside. Image Credit: https://commons.wikimedia.org/wiki/File:Basis_of_Membrane_Potential2.png

21.9.1. Activity: The Sodium Ion Pump

a. What magnitude of work must be done by the cell to move *one* sodium ion from the inside of the cell to the outside?

b. Assume the thickness of the cell membrane is 5 nm. What is the magnitude of the electric field inside the membrane? (Assume the field is uniform.)

c. At rest, the human body uses energy at a rate of approximately 100 W to maintain basic metabolic functions. It has been suggested that 20% of this energy is used just to operate the sodium pumps of the body. Estimate—to one significant figure—the total number of sodium ions that move across cell membranes in the entire body every second.

UNIT 22: INTRODUCTION TO ELECTRIC CIRCUITS

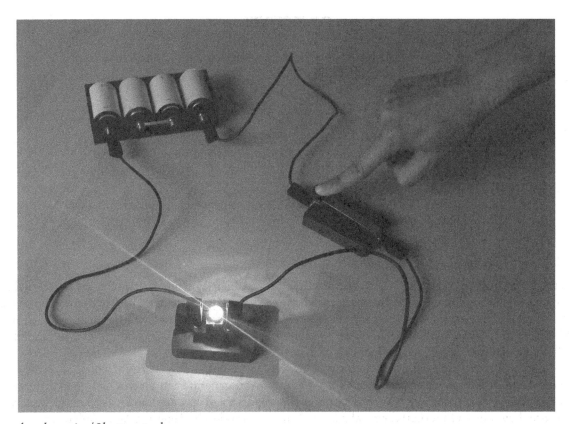

fen deneyim/Shutterstock

Pressing the switch in this simple circuit causes the light bulb to glow. But what physically causes the bulb to light up, and how would changes in the circuit affect the brightness of the bulb? We will investigate these questions and many more in this unit. By the end, you should be able to predict the brightnesses of bulbs in both series and parallel circuits, as well as more complicated arrangements.

UNIT 22: INTRODUCTION TO ELECTRIC CIRCUITS

OBJECTIVES

1. To understand how a potential difference and electric field result in current flow through a conductor.

2. To design, wire, and draw simple circuits using batteries, wires, and switches.

3. To learn to use an ammeter for measuring current and a voltmeter for measuring voltage.

4. To understand how current flows in both series and parallel circuits due to the concept of resistance.

22.1 OVERVIEW

In the following sections we will discover, extend, and apply theories about electric charge, electric fields, and potential difference to the topic of *electric circuits*. The study of circuits will prove to be one of the more practical parts of the introductory physics course, as electric circuits form the backbone of nearly all electronic devices. Without circuits, we wouldn't have electric lights, air conditioners, automobiles, cell phones, computers, or many other common devices.

In the previous unit we used a battery to establish a potential difference (or voltage) across two electrodes to map out the equipotential surfaces. A *battery* generates an electric potential difference using another form of energy. Standard batteries (the type we will use) are known as chemical batteries because they convert (internal) chemical potential energy into electrical energy. As a result of a potential difference, electric charge is repelled from one terminal of the battery and attracted to the other, and this flow of charge can cause a light bulb to glow (among many other things).

In this unit we will explore how charges flow in wires and bulbs and develop models that predict charge flow in both series and parallel circuits. We will also devise ways to test these models using an ammeter to measure the electric current (the rate of flow of electric charge) and a voltmeter to measure the potential difference between two points in the circuit.

TCSWiley/Shutterstock.com   Scanrail1/Shutterstock.com

CURRENT FLOW IN CIRCUITS

Franklin and Electricity

United States Department of
the Treasury/Wikipedia
Commons/CC BY SA 3.0

22.2 WHAT IS ELECTRIC CURRENT?

We have attributed electrical forces between objects to a property of matter
known as charge. Experiments in the late eighteenth and early nineteenth cen-
turies investigated the nature of charge and how apparently different phenomena are all aspects of what is now known as *electricity*. Perhaps the most famous
of these is Benjamin Franklin's "kite experiment." While the specific details are
somewhat in doubt, it does appear that the experiment demonstrated that lightning in the sky is electric in nature. In this unit, we will study how charges in
motion (an electric *current*) form the basis of electric circuits.

Comparing the Effect of a Battery to Rubbing Materials Together

To get at the nature of current, we'll start by comparing the effect of a battery
on a experimental arrangement to that of charge transfer by running materials
together. You'll need the following equipment (the experiments may be done as
a demonstration):

- 2 aluminum plates, approx. 15 cm (e.g., L-brackets)
- 1 conducting, threaded ball (with low mass)
- 1 battery pack or power supply, approximately 300 volts
- 6 alligator clip leads
- 1 black plastic rod
- 1 piece of fur
- 1 glass rod
- 1 polyester cloth
- 1 electroscope (optional)
- 1 Wimshurst generator (optional)

Set up the electroscope (optional) and the metal plates as shown in Fig. 22.1
(as we'll see in a future unit, the metal plates form what is known as a *capacitor*). Separate the metal plates so the gap between them is slightly larger than
the diameter of the conducting ball, and then place the ball carefully between
the metal plates. By "charging" the metal plates, you can test whether or not
different charging methods have different effects on the ball hanging between
the metal plates (and the electroscope). The main charging methods to be tested
are:

1. *Electrostatic Charging by Rubbing:* Charge up a plastic rod by rubbing
   it with fur, and then rub this rod against one of the angle irons. Repeat
   this several times, always touching the same plate. You can also try rubbing the other metal plate with a glass rod that has been rubbed with
   polyester cloth.
2. *Charging with a Battery:* Connect a wire from the negative terminal
   (black) of the battery to one of the plates. At the same time, connect a
   wire from the positive terminal (red) of the battery to the other plate.
3. *Charging with a Wimshurst Generator (optional):* Connect a wire from
   one of the two terminals of the generator to one plate and a wire from the
   other terminal to the other plate.

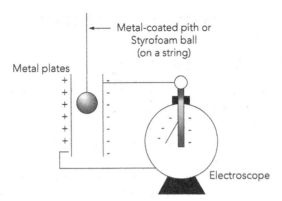

Fig. 22.1. Apparatus for detecting charge and its flow. The connected electroscope is optional.

22.2.1. Activity: Comparing Methods of Moving Charge

a. Use the familiar rubbing method (method #1) to charge both the plates (one negative, one positive). What happens to the metal ball? Optional: what happens to the electroscope? It will probably take multiple swipes with the charged rods to see the effect. Briefly describe what you observe.

b. Next, use the battery (and, optionally, the Wimshurst generator) to "charge" the plates. Describe what you see. What similarities and differences (if any) do you see as compared to charging the plates by rubbing them with the charged rod? Briefly explain why you think the ball acts as it does.

You should have seen that different methods produced essentially the same effect: the conducting ball bounces back and forth between the two plates. Using the battery or Wimshurst generator will likely produce a more significant (and sustained) effect, but it is qualitatively the same as applying charge by rubbing the plastic rod with fur. Both methods involve a *transfer of electric charge*, either by direct contact of the rod with the plate, or through a wire in the case of the battery. This transfer of charge leads us to define an electric current.

The Mathematical Definition of Current

The rate of flow of electric charge is more commonly called *electric current*. If charge is moving, the average current $\langle i \rangle$ is given by

$$\langle i \rangle \equiv \frac{\Delta q}{\Delta t} \quad \text{(average current)} \qquad (22.1)$$

where i is the current and Δq is the amount of charge that flows past a given point in the amount of time Δt.[1] The unit of current is called the ampere (A) and is defined as one coulomb per second: $1 \text{ A} = 1 \text{ C/s}$. The instantaneous current is defined by using a limit to get a derivative:

$$i \equiv \lim_{\Delta t \to 0} \frac{\Delta q}{\Delta t} = \frac{dq}{dt} \quad \text{(instantaneous current)} \qquad (22.2)$$

We note that current is defined as the flow of *positive* charges (even though we now know it is the negatively-charged electrons that move). Because of this definition, we'll see that current flows out of the positive terminal of a battery.

22.3 LIGHTING A BULB

We begin our exploration of circuits and currents by lighting a bulb with a battery. You will need the following equipment for the experiments in this section:

- 1 #14 V bulb
- 1 #14 bulb socket
- 1 D-cell battery, 1.5 V
- 1 D-cell holder
- 2 wires with alligator clip leads, > 10 cm
- common objects: paper clips, pencils, etc.
- magnifying glass (optional)
- 1 SPST (single-pole, single-throw) switch

In the first activity, you will use materials listed above to find arrangements in which the bulb lights (and arrangements in which it does not light). For this activity, keep both the battery and bulb "bare" as in Fig. 22.2.

Fig. 22.2. A possible wiring configuration that might cause a bulb to light in the presence of a battery.

---

[1] Note that our use of the symbol Δ with Δq is slightly different from our past usage. The quantity Δq represents the amount of charge that has flowed past a given point in the amount of time Δt. As we'll see, the total amount of charge in the wire is essentially constant, and so it doesn't really make sense to think of Δq as a *change* in charge.

22.3.1. Activity: What Causes the Bulb to Light Up?

a. Using just the bulb and wires (no other objects yet), describe and/or sketch two different arrangements in which the bulb lights. Think about such things as how the bulb is connected to the battery, reversing direction, etc.

b. Describe and/or sketch two different arrangements in which the bulb does *not* light up.

c. Describe as fully as possible what conditions are needed for the bulb to light.

Conductors and Insulators

In Unit 19 we discussed both *conductors* (materials where charges flow easily) and *insulators* (materials where they don't). Our new, simple circuit allows us to quickly test which materials are conductors and which are insulators.

Put the bulb into the socket and the battery into its holder for easy connections. Use the "alligator clips" at the ends of the wires to connect the battery and bulb and ensure the bulb lights up. Then, disconnect *one* set of alligator clips and insert a variety of objects to see if current flows through each object. Objects you might try include a paper clip, rubber band, pencil and/or pencil lead, key, or paper. Reconnect the alligator clips so that they are connected to different points on the object (but not directly to each other).

22.3.2. Activity: Different Materials in a Circuit

a. List some materials that allow the bulb to light.

b. List some materials that prevent the bulb from lighting.

c. What types of materials seem to be conductors? What types seem to be insulators?

Light Bulbs and Switches

In the final activity of this section, we consider more carefully the design of a light bulb and why it operates the way it does. We also introduce the concept of a *switch*.

Fig. 22.3. Wiring inside an incandescent light bulb.

22.3.3. Activity: Bulb and Switch Function

a. Remove the bulb from its socket and carefully examine both the bulb and socket (use a magnifying glass, if available). Figure 22.3 shows the parts of the bulb that are hidden from view, including the path of the single conducting wire that goes through the bulb. Replace the bulb in its socket and connect the socket to the battery so that the bulb lights up. Now try unscrewing the bulb from its socket—what happens? Briefly explain the basic wiring principle of the bulb and socket based on the results of our earlier activities.[2]

b. Now consider the circuit shown in Fig. 22.4. Predict whether the bulb will light up with the switch open (no contact), closed (contact), neither, or both? Briefly explain why.

---

[2] Our diagram shows a traditional incandescent bulb, which are becoming less common. While other types of bulbs (such as LED and fluorescent) operate using different principles, their basic wiring paths remain the same as an incandescent bulb.

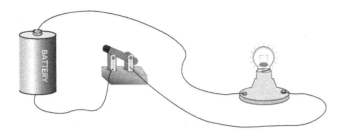

Fig. 22.4. A circuit with a battery, bulb, and a simple switch.

    c. Wire up the circuit in Fig. 22.4 and test your prediction from part (b). What happens?

22.4 CIRCUIT DIAGRAMS

As you might imagine, it can get very time consuming to draw pictures of batteries, bulbs, and switches in circuits. A series of short-hand symbols have been created to represent different elements in a circuit, and these symbols enable one to quickly draw neat, square-looking circuit diagrams. A few symbols for common circuit elements are shown in Fig. 22.5. Notice that the battery has what is known as a *polarity*—one end is the positive side and one end is the negative side. In the circuit diagram, a longer line represents the positive side (indicated with a "+" sign), while the shorter line represents the negative side.

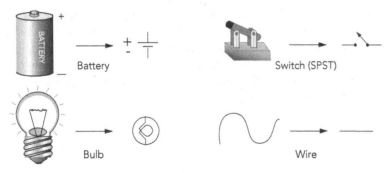

Fig. 22.5. Circuit symbols for a battery, bulb, switch, and wire. The switch is the simplest type known as a single-pole, single-throw (SPST).

    Figure 22.6 shows a simple circuit with both a sketch (left) and a circuit diagram using symbols (right). Compare how the various parts of the diagram correspond to the actual circuit. Before we dig into the details of how circuits work (and to get practice with circuit diagrams), let's apply what you have learned to a simple device. Suppose you want to design a string of lights to hang up around the walls in your room. Your goal in the next activity will be to figure out a way to wire up such a string so that if any one of the bulbs burns out, the other two stay lit. (For simplicity our string of lights has only three bulbs, although an actual set of lights would obviously have more!)

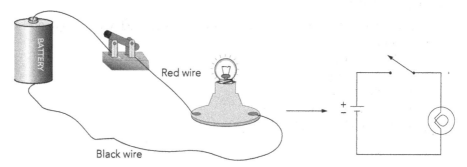

Fig. 22.6. A sketch of a simple circuit and its corresponding diagram.

To complete the activities in this section, you'll need the following equipment:

- 3 #14 bulbs
- 3 #14 bulb sockets
- 1 D-cell battery, 1.5 V
- 1 D-cell holder
- 6 alligator clip leads, > 10 cm
- 1 SPST switch

22.4.1. Activity: Designing a Circuit

Design and construct a circuit with a string of three lights in such a way that if any one of the bulbs burns out, the other two will stay lit. You can simulate a bulb burning out by unscrewing one of the bulbs in its socket (removing a bulb from the socket has the same effect as it burning out). Once you successfully build the circuit, draw the corresponding circuit diagram below.

22.5 A MODEL FOR CURRENT FLOW

We have seen that electric current represents a flow of charges. One could imagine several different ways for current to flow in a circuit. For example, Fig. 22.7 shows four possible models to describe current flow. In the next set of activities, you will design ways to test which of these models is correct. You will need the following equipment:

- 2 ammeters, 0.25 A (either stand-alone or interfaced to a computer)
- 1 #14 bulb
- 1 #14 bulb holder
- 1 D-cell battery, 1.5 V
- 1 D-cell holder
- 1 SPST switch
- 4 alligator clip leads, > 10 cm

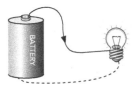

Model A
Current is used up in the bulb, so there will be no electric current left to flow in the bottom wire.

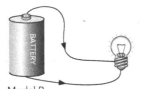

Model B
The electric current will travel in a direction toward the bulb in both wires.

Model C
The direction of the current will be in the direction shown, but there will be less current in the return wire.

Model D
The direction of the current will be as shown, and it will be the same in both wires.

Fig. 22.7. Four alternative models for current flow.

22.5.1. Activity: Predicting a Model for Current Flow

Which diagram in Fig. 22.7 do you think best describes how current flows in the simple circuit shown? Why? Talk it over with your partners and make a prediction.

Current measurements

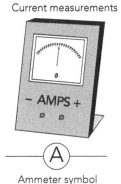

Ammeter symbol

Fig. 22.8. An ammeter and its symbol.

To test the models, we need a method of measuring current (both the magnitude and direction). Current is measured in amperes ("amps" for short), and a device that measures current is called an *ammeter*. It turns out that one ampere is quite a large current, and so you will often see current measured and reported in milliamperes (mA), or 1/1000 of an ampere. On a circuit diagram, an ammeter is represented as a circle with an "A" in the middle.

Ammeters come in many forms. An analog version is shown in Fig. 22.8, where a dial indicator moves back and forth to show the current (either positive or negative). There are also digital meters, in which a digital display shows the current and its sign (see Fig. 22.9). Digital meters can typically measure more than just current and are therefore called *multimeters*. Finally, one can measure current using a laboratory interface connected to a computer. Any of these forms will suffice for this activity.

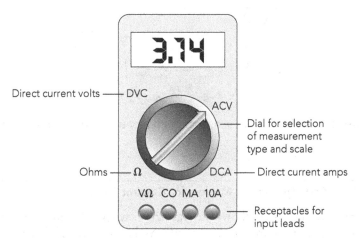

Fig. 22.9. Diagram of a multimeter that can be used to measure resistances, currents, and voltages.

The digital multimeter is a very common scientific device, so we briefly describe its use here. A multimeter is typically used to measure either current or potential difference (voltage), although most multimeters can measure other quantities as well. For example, the multimeter of Fig. 22.9 has settings for measuring direct current (DCA), direct current voltage (DCV), alternating current voltage (ACV), and resistance (Ω). There is one negative connection, or *terminal*, typically labeled COM or CO to represent the "common" terminal; the black lead goes into this terminal. There are multiple positive terminals, and the one you choose is based on what you wish to measure (the red lead goes into the appropriate positive terminal). The multimeter in Fig. 22.9 has three positive input terminals: 10A for measuring large currents, MA for measuring small (milliamp-scale) currents, and VΩ for measuring either voltage or resistance.

You will be measuring current in the next activity, and just to be safe you should start with the "10A" terminal. A positive current reading means that current is flowing into the positive terminal and out of the negative terminal, while a negative current reading means that current is flowing in the opposite direction. Note that you will need to use both the magnitude and sign of the current to distinguish between the models in Fig. 22.7!

Figure 22.10 shows how to connect an ammeter to measure current in a circuit. The ammeter measures the current flowing through one part of a circuit, and you must "insert" the ammeter at the point of interest. In other words, you must *disconnect* wires in the circuit, put in the ammeter at the point you want to measure the current, and then reconnect the wires to the ammeter. For example, the ammeter shown in Fig. 22.10 will measure the current in the bottom wire (coming out of the left side of the bulb and going into the negative terminal of the battery).

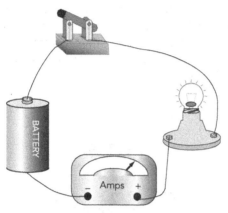

Fig. 22.10. Measuring current in a circuit with a battery, bulb, and ammeter (and switch).

22.5.2. Activity: Testing Models for Current Flow

a. Set up the circuit shown in Fig. 22.10, where the ammeter measures the current coming *out of the bulb*. Make sure the light bulb is glowing and that you have connected the positive and negative terminals of the ammeter as shown. If you are having trouble figuring out how to wire the circuit, you might find it useful to start at the positive terminal of the battery and "walk" your way around the circuit. For example, the positive terminal of the battery is connected by a wire to one side of the switch; the other side of the switch is connected by a wire to one side of the light bulb socket; and so on. Once you have it working, write down the magnitude of the current at this point in the circuit. Is the current positive or negative?

b. If you wired the circuit correctly, your result for part (a) should have been a positive current. Try reversing the ammeter in your circuit. What do you find? Switch the ammeter back when you are done.

c. Next, add a second ammeter to the circuit to simultaneously measure the current going *into the light bulb*. You will need to disconnect the wire going from the switch to the light bulb and insert the ammeter between them. Think carefully about how you put the ammeter into the circuit so that you can distinguish the direction of current! Once you have the circuit working again, write down the magnitude and sign of the current at this point in the circuit.

d. Based on your measurements, which model of Fig. 22.7 best describes how current flows in this circuit? Briefly explain how your experimental observations support your conclusion.

You should have found that the current before the bulb has the same magnitude and direction as the current after the bulb, making Model D the only one consistent with our results. Notice that current is not "used up" in the bulb; the same amount of current that flows into the bulb also flows out of the bulb. In fact, this same current also flows through all other elements in the simple circuit of Fig. 22.10, including the wires, the battery, and the switch.

People sometimes find this surprising, as it seems like the bulb must be "using energy" when it is on. If so, how can the current be the same both before and after the bulb? The answer to this puzzle is that the energy (or power) dissipated in a bulb cannot depend only on the current! In the next section, we'll examine this missing piece.

CURRENT AND POTENTIAL DIFFERENCE IN CIRCUITS

22.6 MEASURING POTENTIAL DIFFERENCE

As we will see, a battery is a device that maintains a fixed potential difference *across* its terminals, and this means it is capable of supplying energy to charges that move in a circuit. The relationship between the potential differences in a circuit and the currents that flow is an essential part of developing an understanding of electric circuits. Because potential differences are measured in volts, a potential difference is often referred to as a *voltage*. Although we will use these two terms interchangeably, remember that when we say "voltage," we are still referring to a *difference*—a voltage difference—between two points in a circuit.

Figures 22.11 and 22.12 show a circuit with a battery, bulb, and switch. The diagrams also show a *voltmeter*, which is used to measure the potential difference across any element (or combination of elements). The voltmeter measures the potential different across the battery in Fig. 22.11 and across the bulb in Fig. 22.12. The word *across* is descriptive of how the voltmeter must be connected to measure voltage. You don't need to "break" the circuit to make a measurement like you do with an ammeter. Instead, just touch the two leads of the voltmeter to the two locations in the circuit, and the meter tells you the potential difference between these two points (positive terminal minus the negative terminal). Note that this is different than how we used an ammeter to measure the current *through* an element!

To do the activity in this section, you will need:

- 1 D-cell battery, 1.5 V
- 1 D-cell holder
- 1 #14 bulb
- 1 #14 bulb holder
- 1 SPST switch
- 1 voltmeter

Note: If you are using a multimeter, be sure to set the dial for reading DCV (Direct Current Voltage).

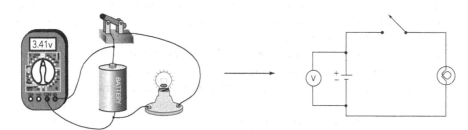

Fig. 22.11. Illustration of a circuit (and circuit diagram) of a voltmeter used to measure the voltage across a battery in a simple circuit with a battery, a bulb, and a switch. The voltmeter is represented in the circuit diagram as a "V" with a circle around it.

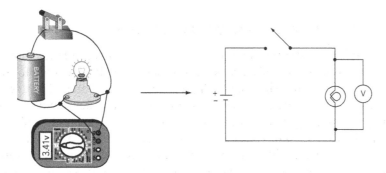

Fig. 22.12. Illustration of a circuit (and circuit diagram) of a voltmeter used to measure the voltage across a bulb in a simple circuit with a battery, a bulb, and a switch.

22.6.1. Activity: Voltage Measurements and Non-Ideal Elements

a. Wire up the circuit (without the voltmeter) and close the switch so that the bulb lights up. Then, start by connecting both the positive and negative leads of the voltmeter to the *same point* in the circuit (i.e., both leads are touching the exact same location). Observe the reading. Repeat this process at another point in the circuit. What do you conclude about the potential difference when the leads are connected to the same point? Open the switch when you are done.

b. Next, make a *prediction* about the voltage across the battery compared to the voltage across the bulb when the switch is closed. Explain your prediction.

c. Close the switch and test your prediction by using the voltmeter to measure the voltage *across* the battery. Touch the red lead to the positive terminal of the battery and the black lead to the negative terminal (if your battery holder does not allow you to access the terminals, you can connect the voltmeter to the alligator-clip leads coming from the battery). What happens if you reverse the leads (red to negative terminal and black to positive terminal)?

d. Now use the voltmeter to measure the voltage *across* the bulb. Touch the red lead to the side of the bulb closest to the positive terminal of the battery and the black lead to the side of bulb closest to the negative terminal of the battery. How do the voltage across the battery and the voltage across the bulb compare when the bulb is lit?

e. Hopefully, you saw that the voltage across the battery and the bulb were roughly the same when the bulb is on. But they may not have been *exactly* the same. To get at why this is, use the voltmeter to measure the potential difference across one of the wires (it doesn't matter which one). Touch the red lead to the clip on one side of a wire and the black lead to the clip on the other side of the same wire (with no circuit element in between). What voltage do you measure across a wire when the bulb is on? **Note**: You must connect the voltmeter to the metal leads at the ends of the wire, since the wire itself is "insulated."

You may have seen a small but non-zero voltage across a wire (if your wire is very "good," the voltage will read essentially zero). An ideal wire should not have any voltage across it, no matter how much current is flowing. As we'll see in upcoming activities, this has to do with the amount of resistance in the wire as compared to a bulb or other elements. However, no circuit element is perfect, and you'll need to keep this in mind as we continue. Let's do one more measurement that demonstrates how a real circuit element does not behave in an ideal manner.

f. Return the voltmeter to measure the potential difference across the *battery*. While monitoring the voltage across the battery, open and close the switch a couple of times. What do you notice about the voltage across the battery when the bulb is on compared to when the bulb is off?

22.7 CURRENT, VOLTAGE, AND MORE BULBS

Now that we know how to measure both the current *through* a circuit element and the voltage *across* an element, we can begin to explore the relationship between these two quantities as we add more bulbs to a circuit. To perform the next few activities, you will need:

- 1 D-cell battery, 1.5 V
- 1 D-cell holder
- 2 #14 bulbs (with the same brightness)
- 2 #14 bulb holders
- 1 SPST switch
- 1 voltmeter
- 1 ammeter, 0.25 A

Connect the voltmeter and ammeter so that you are measuring the voltage *across* the battery and the current *through* the circuit at the same time (see Fig. 22.13). **Reminder**: For this simple circuit, Model D for current flow tells us that the current through the bulb, both wires, and the battery must be the same everywhere in the circuit. Therefore, we only need to measure the current at one location.

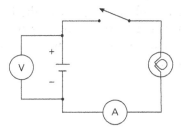

Fig. 22.13. A single-bulb circuit diagram showing connections to measure the voltage across the battery and the current through the circuit.

22.7.1. Activity: Current and Voltage for a Single Bulb

a. Construct the circuit in Fig. 22.13. Measure the voltage across the battery when the switch is closed and the light is on.

Voltage across the battery (with one bulb): _____ Volts

b. Measure the current through the circuit when the switch is closed and the light is on. Because this same current flows *through* the battery in this single-loop circuit, we'll refer to the current as "coming from the battery."

Current coming from the battery (with one bulb): _____ Amps

Now suppose we connect a second bulb in the circuit as shown in Figure 22.14. We are interested it what changes due to this additional bulb. We'll start with some predictions.

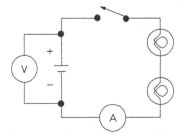

Fig. 22.14. Two bulbs in a row connected to a battery with a voltmeter and ammeter set up to measure voltage across the battery and current through the circuit.

22.7.2. Activity: Current and Voltage for Two Bulbs in a Row

a. For the circuit shown in Fig. 22.14, *predict* how the voltage across the battery will compare to that with only one bulb (Fig. 22.13). Will it change significantly? Briefly explain your prediction.

b. *Predict* what you think will happen to the brightness of the bulb in Fig. 22.13 when you add a second bulb as shown in Fig. 22.14. Will it get brighter, dimmer, or remain the same as compared to before? Briefly explain your prediction.

c. *Predict* how the current will compare for the circuits shown in Figs. 22.13 and 22.14. When the second bulb is added, will the current increase, decrease, or remain the same as compared to before? Briefly explain your prediction.

d. Now go ahead and add the second bulb to the circuit and test your predictions. Make sure to add the bulb as shown in the circuit diagram of Fig. 22.14. Measure the voltage across the battery with the switch closed. Did it change *significantly* as compared to the circuit with a single bulb?

Voltage across the battery (with two bulbs): _____ Volts

e. Did the first bulb change brightness when you added the second bulb to the circuit? What about the current through the circuit—did it change when the second bulb was added?

Current coming from the battery (with two bulbs): _____ Amps

f. Summarize how the addition of the second bulb in Fig. 22.14 affected the voltage across the battery and the current through the battery. Why do you think this happens? (The next activity will help us understand this.)

g. Based on your results, does the battery appear to be a source of constant voltage or constant current?

22.8 A MODEL FOR VOLTAGE AND CURRENT

The fact that current is not "used up" in passing through a bulb seems counter-intuitive to many people. It can be helpful to construct a model of a gravitational system that is in some ways analogous to the electrical system we are studying. Before doing so, we should briefly describe what is happening electrically!

Imagine connecting a battery to a circuit that includes a wire and a bulb. For the purposes of our model, it's best to talk about the *electrons* as moving through the circuit (as is actually the case). We have seen that a battery provides a constant source of (electric) potential difference ΔV between its two terminals. This means that the (electrical) potential energy is higher for the electrons that are already present at the negative terminal of the battery (and along the wire attached to the negative terminal). The potential difference of the battery also produces an electric field in the wires and bulb that points along the wires from the positive terminal of the battery to the negative terminal. Because of this electric field, the existing electrons are accelerated through the wire (and bulb), resulting in an increase in their kinetic energy (and corresponding decrease in their electrical potential energy). However, the electrons collide with and scatter (bounce) off the atoms in the bulb filament. Each collision slows the electron down, but the electron then accelerates again until it collides with another atom.

Figure 22.15 provides a sketch of the process, with an electron staggering its way through the material of the filament. Although the direction of the electron's instantaneous velocity is continually changing, it will have some average velocity $\langle \vec{v} \rangle$ as it moves through the material. This average velocity is often referred to as the *drift velocity*.

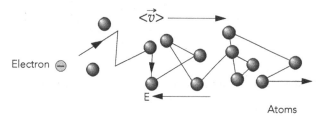

Fig. 22.15. An electron in a uniform electric field staggering through a material as a result of collisions with atoms, giving it an average drift velocity $\langle \vec{v} \rangle$.

The *resistance* to flow the electrons experience depends on the material the wire (or bulb filament) is made of and the geometry of the wire/filament. For example, a thick wire made of copper has a low resistance to current flow—there tend to be fewer collisions and the electrons move relatively freely. On the other hand, a light bulb with a very thin filament made of tungsten has a much higher resistance to current flow—there are frequent collisions that repeatedly slow down the electrons. The current, measured in amperes (coulombs per second), is essentially an indication of how quickly the electrons can move through the material; more resistance results in a slower drift velocity and a lower current.

Building a Working Model for Current Flow

A typical solid has a three-dimensional crystalline structure, an example of which is shown in Fig. 22.16. It is through this structure that the electrons move. To simplify the situation, we will consider a two-dimensional mechanical analog to model our picture of current flow through conductors (see Fig. 22.17). The instructions below provide a list of required items and an overview of how to build the model. **Note**: Your instructor may already have a working model built; if so, you can skip the paragraph describing how to build it.

- Collection of identical steel balls or marbles
- 1 ramp (with adjustable angle)
- 1 ruler

- 1 protractor
- 1 board with face-centered pegs or
  - 1 polyurethane insulation board, approx. 6″ × 18″
  - 2 poster boards, approx. 2″ × 18″
  - 100 plastic push pins
  - 2 graph paper sheets, 8.5″ × 11″ with 1/4″ grid lines
  - 1 stopwatch

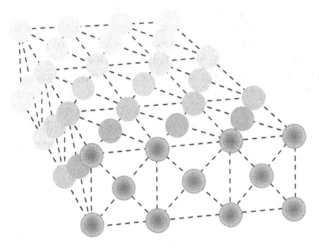

Fig. 22.16. Arrangement of atoms in a face-centered cubic crystal.

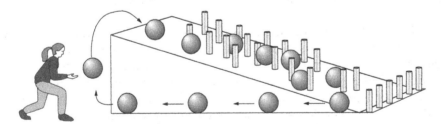

Fig. 22.17. A mechanical analog for electrical current flow. Ideally, a hopper of balls is located at the top of the ramp to serve as a "reservoir" of marbles.

To build the model, you can take an adjustable wooden ramp and mount a piece of insulation board on it. Push pins can be poked into the soft insulation board to model atoms. We'd recommend placing graph paper on top of the insulation board to facilitate equal spacing of the atoms, and the push pins should represent a face-centered arrangement. A three-dimensional face-centered arrangement of atoms is shown in Fig. 22.16; you'll want to represent this in two dimensions. The maximum spacing between corner atoms should be large enough to allow balls to "flow," but small enough to force regular collisions.

And don't forget to put an atom in the center of each set of four corner atoms. Strips of poster board can be pinned into the sides of the insulation board to make side rails, and a collection of marbles or steel ball bearings flow through the circuit.

In Activity 22.8.1 we will investigate the analogy between the gravitational and electrical situations. As depicted in Fig. 22.17, the pegboard ramp is propped up so that one end is a height h above the other end so that the steel balls starting at the top run down the pegboard. We can imagine a person (not to scale!) lifting the balls back up to the top of the ramp so they can repeat the journey. Ideally, there would be hopper containing a sufficient number of balls so that there is always a reservoir available, with the person simply refilling the hopper as the balls arrive at the bottom. **Note**: While our mechanical analogy is *not* a perfect model for current flow, it will hopefully help you understand some of the basic processes at work in a circuit.

Place a collection of the balls at the top (or open the hopper) and let them roll down the ramp while running into the pegs. For the most realistic connection to the electrical case, there should always be a supply of balls at the top of the pegboard. After you've tested the model experimentally, answer the questions in Activity 22.8.1.

22.8.1. Activity: Explaining the Features of the Model

a. What do you think will happen to the ball current (i.e., the rate of ball flow) if fewer pegs are placed in the path of the balls?

b. What do you expect will happen to the ball current if the ramp were raised to a greater height so that balls have more gravitational potential energy when they start?

c. Examine the list below and draw lines between the elements of a circuit consisting of a battery, wire, and bulb and the elements of the corresponding model.

Electrical	Mechanical Analogy
Resistance to flow	Average speed of balls
Battery "action"[3]	Number of pegs encountered
Current	Person raising the balls
Voltage of battery	Height of the ramp

---

[3] For chemical batteries like ours, this is supplied by a chemical reaction that takes place inside the battery and provides electrical potential energy to the charge carriers.

d. What should happen to the gravitational potential energy of the balls as they roll down the ramp? Where does this energy ultimately go? Similarly, what do you think happens to the electrical potential energy of an electron as it passes through the filament of a bulb and undergoes lots of collisions?

e. Discuss how our model shows that (1) current is the same everywhere in a single-loop circuit, and (2) electric current isn't "used up" when it flows through a bulb.

f. Raising the height of the ramp increases the average speed of the balls down the ramp. By analogy, what should you do in our simple circuit if you want to increase the current through the bulb (assuming the type of bulb stays the same)?

g. To model two bulbs in a row (like in Activity 22.7.2), we could add an additional pegboard to the experiment such that the start of the second pegboard is attached to the end of the first one (making the combined pegboard twice as long). The balls would begin at the same initial height but must travel through both pegboards before reaching the tabletop. How would this modification affect the ball current in our mechanical experiment? Does this make sense based on the results of Activity 22.7.2? Briefly explain.

We just discussed the model in terms of the negatively-charged electrons moving through the material (which is what actually happens in a real circuit). But since current in defined in terms of positive charge carriers, let's briefly summarize the situation from that point of view. Positive charge carriers on the positive side of the battery have a higher potential (and thus a higher potential energy) than those on the negative side, and this causes the charges to flow "downhill" in the circuit (toward lower potential energy). Alternatively, one can think about the potential of the battery providing an electric field in the wires and bulb, and this electric field imparts a force on the charge carriers. The resistance to the flow of charge in the bulb is analogous to the mechanical resistance offered by the pegs on the pegboard in our gravitational model. Even though all the current returns to the battery after flowing through the bulb, the charge carriers have lost energy due to the collisions with the atoms in the bulb filament; this energy ends up as heat and light.

22.9 SERIES AND PARALLEL CIRCUITS

We observed a number of results over the last several activities, including that a bulb lights up when electric current flows through it, that current only flows when it has a complete, conducting path from the positive terminal of the battery to the negative terminal, and that current in a simple, single-loop circuit is the same through all elements. We also found that a battery supplies essentially the same voltage whether it is connected to one light bulb or two, but putting two light bulbs in a row affects the current because two bulbs (in a row) offer more resistance to the flow of current than a single bulb.

In the next several activities, we will examine more complicated circuits and compare the currents through different parts of the circuits by comparing the brightness of the bulbs and/or measuring the current with an ammeter. To make the required observations, you should wire circuits with a fresh battery and verify that all your bulbs have the same brightness when separately connected to the battery. You'll need the following equipment:

- 1 D-cell battery 1.5 V
- 1 D-cell holder
- 6 alligator clip leads, > 10 cm
- 3 #14 bulbs (with identical brightness on own)
- 3 #14 bulb holders
- 1 SPST switch

Hint: Helpful symbols for reporting relative brightness are

> "greater than"

< "less than"

= "equal to"

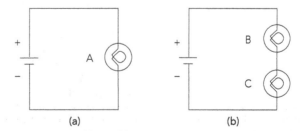

Fig. 22.18. Two different circuits with identical components: (a) battery with a single bulb, and (b) battery with two bulbs in a row.

22.9.1. Activity: Review of a Series Circuit

a. Let's start with a quick review of our observations from Section 22.7. Consider the two circuits shown in Fig. 22.18. Rank the relative brightness of the three bulbs from brightest to dimmest. If two or more bulbs are equal in brightness, indicate this in your response. Briefly explain the reasons for your rankings. If you wish (or if there is disagreement), hook up the circuits and observe the brightnesses again.

b. If we think of each bulb as providing a resistance to the flow of current, how do you think the *total resistance* to the flow of current through the circuit will be affected by the addition of more bulbs (in a row) in Fig. 22.18(b)?

c. Therefore, what should happen to the *total current* through the circuit due the addition of more bulbs in a row?[4]

When bulbs are added along a single path, as in the previous activity, we say they are wired *in series*. Another way to connect bulbs (or other elements) is along separate paths, which is known as being wired *in parallel*. The precise definitions of series connections and parallel connections are summarized in Fig. 22.19.

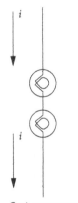

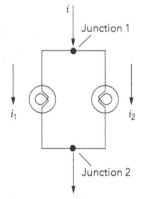

Series connection:
Two bulbs are in series if all the current that flows through one bulb must also flow through the second bulb.

Parallel connection:
Two bulbs are in parallel if their terminals are connected so that at each junction one terminal of one bulb is directly connected to one terminal of the other.

Fig. 22.19. Series (left) and parallel (right) connections.

---

[4] While we can correctly predict an increase or decrease in current as more bulbs are added in a row, it is hard to be *quantitatively* correct. This is because the resistance of a bulb to current flow changes as the current through the bulb changes (and the bulb changes temperature). Later, we will find better quantitative agreement using circuit elements called *resistors* that have a constant resistance regardless of the current.

In the next activity we'll consider a circuit with two bulbs wired in parallel, such as in Fig. 22.20. Note that even though the bulbs in Fig. 20.20 are drawn differently than the parallel arrangement in Fig. 22.19, it follows the same rule: the "tops" of the two bulbs are directly connected by wire, as are the "bottoms" of the bulbs. Understanding why these connections are the same is good practice, as you will need to be able to determine when circuits are *electrically equivalent* even if they don't look exactly the same. When doing this, it is helpful to remember that the voltage does not change along ideal wires, so the exact shape of the wires and layout is not important. What *is* important are the available paths the current can take.

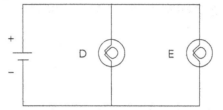

Fig. 22.20. A parallel circuit of two bulbs.

As with the series circuit, you'll want to compare two bulbs in parallel to a single-bulb circuit. Instead of building two separate circuits, inserting a switch as shown in Figure 22.21(b) allows you to quickly go back and forth between a single-bulb circuit and two bulbs in parallel. With the switch open, the path to bulb E is broken, and no current will flow through bulb E (the circuit is simply a single bulb connected to a battery). With the switch closed, you have two bulbs in parallel.

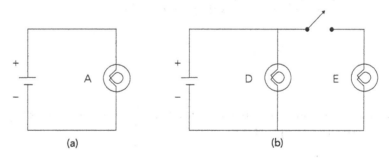

Fig. 22.21. When the switch is open in circuit (b), only bulb D is connected to the battery, so it acts like the circuit shown in (a). When the switch is closed, bulbs D and E are connected to the battery in parallel.

22.9.2. Activity: Relative Currents in a Parallel Circuit

a. *Before* building the circuit, *predict* the relative brightnesses of bulbs A, D, and E in Fig. 22.21 with the switch closed (two bulbs in parallel). Rank the bulbs from brightest to dimmest using >, <, and =. Briefly explain the reasons for your prediction.

b. Do you think opening and closing the switch in Fig. 22.21(b) will affect the current *through bulb D*? Briefly explain.

c. Now wire up the circuit shown in Fig. 22.21(b) and, by opening and closing the switch, observe the relative brightnesses of bulbs A, D, and E. Write down the ranking. How did your observations compare to your predictions? **Note**: Because your battery is not ideal, you may observe small changes in bulb brightness when the switch is closed (only significant changes should be noted).

d. Does opening and closing the switch appear to *significantly* affect the current through bulb D? Explain how you know.

e. Based on your observations, how do the relative currents between bulbs A, D, and E compare?

You should have seen that two bulbs in parallel have equal brightness to each other, and equal brightness to a single bulb connected to the same battery.[5] Note that this is *not* what happens when we wire two bulbs in series. In the series situation, the two bulbs are significantly dimmer than a single bulb alone, indicating less current is flowing through them. In the next set of activities, we will explore these circuits in more detail and consider voltages as well as currents.

VOLTAGE AND CURRENT IN MORE COMPLEX CIRCUITS

22.10 BATTERY VOLTAGE AND CURRENT IN PARALLEL CIRCUITS

In Activity 22.9.2 we saw that two bulbs in parallel had the same brightness as each other, which was the same as a single bulb connected to the same battery. In other words, when the switch is closed in Fig. 22.21, the current through bulbs A, D, and E are the same (when the switch is open the current through bulbs A and D are the same and there is no current through bulb E). We would now like to think about the circuit as a whole. In particular, is the current through the battery always the same no matter what is connected, or does it change depending on the circuit? What about the voltage? We will answer these questions in this section.

To complete the activities in this section, you will need the following:

- 1 D-cell battery, 1.5 V
- 1 D-cell holder
- 2 #14 bulbs (with identical brightness on own)
- 2 #14 bulb holders
- 6 alligator clip leads, > 10 cm
- 2 ammeters, 0.25 A
- 1 voltmeter
- 1 SPST switch

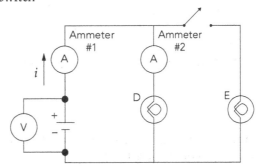

Fig. 22.22. In this circuit, ammeter #1 is connected to measure the current through the battery, while ammeter #2 measures the current through bulb D. A voltmeter measures the voltage across the battery.

---

[5] You may have noticed that the two bulbs in parallel are each *slightly* dimmer than the single bulb. This small difference is the result of a non-ideal battery, whose voltage drops slightly when the current increases.

22.10.1. Activity: Predicting Battery Current and Voltage

a. How do you think closing the switch in Fig. 22.22 will affect the *current through* the battery—that is, the current flowing through ammeter #1? Briefly explain why.

b. How do you think closing the switch in Fig. 22.22 will affect the *voltage across* the battery? Briefly explain why.

To test your predictions, construct the circuit in Fig. 22.22 by placing the voltmeter *across* the battery and inserting the ammeters into the circuit as shown. **Note**: As always, make sure that bulbs D and E have roughly the *same* brightness when connected to the battery alone.

22.10.2. Activity: Observing Battery Voltage and Current

a. Measure the *current through* the battery (ammeter #1) as you open and close the switch. Record your results below.

Battery current with switch open: _____ Amps

Battery current with switch closed: _____ Amps

b. Measure the *current through* bulb D (ammeter #2) as you open and close the switch. Record your results below.

Bulb D current with switch open: _____ Amps

Bulb D current with switch closed: _____ Amps

c. Finally, measure the *voltage across* the battery as you open and close the switch. Record your results below.

Battery voltage with switch open: _____ Volts

Battery voltage with switch closed: _____ Volts

d. Use your observations to predict how you think the current through a battery (the "total circuit current") will change as you increase the number of bulbs connected *in parallel*. Briefly explain.

e. Compare this prediction to what was discussed in Activity 22.9.1 relating the current through the battery to the total *resistance* of the circuit. Does adding more bulbs in parallel increase, decrease, or not change the *total resistance* of the circuit?

f. Explain your answers to parts (d) and (e) in terms of the number of paths available in the circuit for current to flow through.

g. You have now looked at wiring two bulbs in series and in parallel. What can you say about the *total resistance of the circuit* when you wire bulbs in series versus in parallel? Does the total resistance to current flow (and therefore the current through the battery) depend on how the bulbs are wired?

22.11   MORE COMPLEX CIRCUITS

We will quantify the relationship between voltage and current in the next unit. But it turns out that one can make qualitative predictions about circuits using arguments related to current alone. To finish out this unit we examine more complex circuits using the tools we've developed. For the activities in this section, you will need the following:

- 4 #14 bulbs (with identical brightness on own)
- 4 #14 bulb holders

- 1 SPST switch
- 2 ammeters, 0.25 A (optional)
- 1 voltmeter (optional)
- 3 D-cell batteries, 1.5 V
- 3 D-cell holder

OR

- 1 lantern battery, 6.0 V

Applying Your Knowledge to a More Complex Circuit

Consider the circuit consisting of a battery and two bulbs in series as shown in Fig. 22.23(a). We want to consider what happens if you add a third bulb in parallel with bulb C, as shown in Fig. 22.23(b). **Note**: You can either use a 6V lantern battery or three D-cell batteries connected in series in the 3 D-cell holder. We will discuss the details of connecting batteries in series and parallel in the next unit.

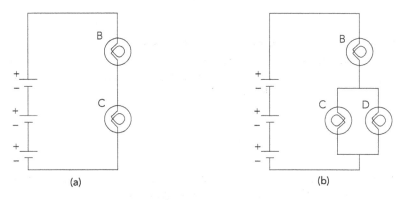

Fig. 22.23. Two different circuits (all bulbs are identical).

22.11.1. Activity: Predictions for a Complex Circuit

a. In Fig. 22.23(b), bulbs C and D are in parallel with each other. But what about bulb B? How would you describe bulb B in relation to bulbs C and D? (You may need to review the definitions of series and parallel connections given earlier in this unit.)

b. Consider the change that takes place between panels (a) and (b) of Fig. 22.23. Is the resistance of the parallel combination of C and D larger than, smaller than, or the same as the resistance of bulb C alone in panel (a)? Briefly explain.

c. Using your answer to part (b), determine whether the resistance of the combination of bulbs B, C, and D in panel (b) is larger than, smaller than, or the same as the combination of bulbs B and C in panel (a)? Briefly explain.

d. Use your responses thus far to predict how the current through bulb B will change, if at all, when bulb D is added in parallel to bulb C (that is, going from panel (a) to panel (b)). In other words, should bulb B get brighter, dimmer, or stay the same when bulb D is added? Briefly explain.

e. Finally, predict the brightness ranking for all the bulbs after bulb D has been added to the circuit. Briefly explain.

It's now time to construct the circuit. Note that as in Section 22.10, we can use a switch to quickly change between the two circuits shown in Fig. 22.23. Convince yourself that the circuit in Fig. 22.24(a) is identical to Fig. 22.23(a) when the switch is open (as shown), and identical to Fig. 22.23(b) when the switch is closed. **Optional**: You can add a voltmeter and two ammeters to the circuit as shown in Fig. 22.24(b), allowing you to make quantitative measurements. If you don't use the meters, the brightnesses of the bulbs will allow you to determine if the current through them has changed.

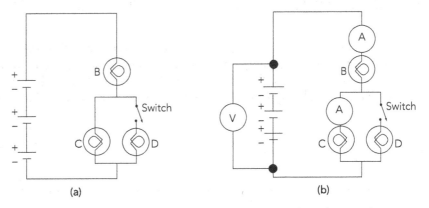

(a)   (b)

Fig. 22.24. (a) This circuit is equivalent to Fig. 22.23(a) when the switch is open, and to Fig. 22.23(b) when the switch is closed. (b) The same circuit with ammeters and a voltmeter connected to measure the current through bulb B, the current through bulb C, and the voltage across the battery.

22.11.2. Activity: Observing a Complex Circuit

a. Set up the circuit shown in Fig. 22.24(a) and observe the brightness of bulbs B and C when the switch is open, and the brightness of all three bulbs when the switch is closed (adding bulb D in parallel with bulb C). What happens to the brightness of bulb B when bulb D is added in parallel with bulb C? Is this what you predicted?

b. Rank the brightness of bulbs B, C, and D with the switch closed. Is this what you predicted?

c. If your observations do not agree with your predictions, what changes do you need to make in your reasoning?

d. **Optional**: Use the voltmeter and two ammeters to measure the voltage across the batteries, the current through bulb B, and the current through bulb C when the switch is opened and closed. Record your measurements below.

Measurement	Before switch closed	After switch closed
Combined battery voltage	Volts	Volts
Bulb B current	Amps	Amps
Bulb C current	Amps	Amps

e. Based on the change in brightness you observed for bulb B (or your measured current through bulb B), what can you say happens to the *total current through the battery* when the switch is closed? Thus, what do you conclude happens to the *total resistance* in the circuit? Explain.

You should have seen that bulb B gets *brighter* when the switch is closed. Therefore, the current through bulb B has increased, and since all the current from the battery must go through bulb B, we can conclude that the total current through the battery must also increase. In other words, adding bulb D in parallel with bulb C *decreases* the total resistance of the circuit, consistent with our earlier reasoning that adding bulbs in parallel adds more possible paths for current, decreasing the overall resistance.

You should also have seen that bulb C gets *dimmer* when the switch is closed. This result is a little harder to explain at this point, so we will wait to discuss it until the next unit. However, if you think about it carefully, the fact that bulb C gets dimmer while the total current goes up tells you something about how much the total current must have increased.

Series and Parallel Branches

Let's look at a somewhat more complicated circuit to see how series and parallel parts of a complex circuit affect one another. The circuit shown in Fig. 22.25 has two "branches": branch 1 contains only bulb A while branch 2 contains bulbs B, C, and D. As usual, we'll start with predictions before making observations. **Note**: Do *not* disconnect your existing circuit! You can simply add bulb A to your circuit, making sure that it is connected as the circuit diagram indicates (please ask if you have questions).

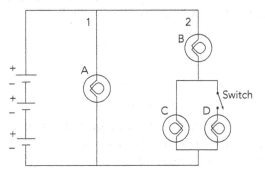

Fig. 22.25. A complex circuit with series and parallel connections. Branch 1 contains only bulb A while branch 2 contains bulbs B, C, and D.

22.11.3. Activity: Series and Parallel Branches

a. When the switch in Fig. 22.25 is *open*, are any bulbs connected in *series* with each other? (If needed, review the definitions of series and parallel connections given earlier in this unit.)

b. When the switch is *open,* are any bulbs connected in *parallel* with each other? Briefly explain.

c. When the switch is *closed,* are any bulbs connected in *series* with each other? Briefly explain.

d. When the switch is *closed*, are any bulbs connected in *parallel* with each other? Briefly explain.

e. What do you predict will happen to the current in branch 1 for each of the following alterations in branch 2: (*i*) opening and closing the switch, and (*ii*) unscrewing bulb B.

f. What do you predict will happen to the current in branch 2 if you were to unscrew bulb A in branch 1.

g. Connect the circuit in Fig. 22.25 and observe the effect of each of the alterations in parts (e) and (f). (Replace bulb B before you test part (f).) Remember that you can use the brightness of each bulb as an indication of any change in current flowing through it. Record your observations for all three cases.

h. Compare your results with your predictions. If needed, account for any differences between your predictions and observations.

i. This circuit has two separate, parallel branches connected across the battery. What can you conclude about how changes in one branch affect other (parallel) branches?

22.12   PROBLEM SOLVING

A More Complicated Switch

In household wiring, you have probably encountered what is known as a "three-way switch" circuit. This configuration allows a single light bulb (or multiple bulbs) to be turned on or off using two different switches. For example, the switches could be located at opposite ends of a large room, or at the top and bottom of a stairway. The standard set-up involves the following: *two* single-pole, double-throw switches (SPDT), one light bulb, a power supply (a battery in our case), and wires.

A SPDT switch has a central terminal that can be connected to one of two other terminals. Unlike a SPST switch that can be "open" or "closed," a SPDT switch has the settings "closed side 1" and "closed side 2" (technically it also can be "open," when the central terminal is not connected to either side, but this is not used in household wiring applications). A drawing of a SPDT switch and its circuit symbol are shown in Fig. 22.26.

Single-pole double-throw switch (SPDT)

Fig. 22.26. A single-pole, double-throw switch (SPDT), which has a middle terminal that can be connected to one of two side terminals.

22.12.1. Activity: A Three-Way Switch

Design and construct a circuit with two SPDT switches that allows you to turn a single bulb on or off using *either* of two switches. Both switches can only go back and forth between the two "closed" configurations (you can't leave a switch in the "open" position). No matter whether the bulb is on or off, either switch should be able to change its state (on → off or off → on). Test out your ideas and draw your final circuit diagram below. Briefly explain how your circuit operates.

UNIT 23: CIRCUIT ANALYSIS

Martin Valigursky/Shutterstock

As a result of atmospheric processes, negative charges build up at the base of clouds causing a potential difference between them and the ground beneath. During lightning strikes, this charge difference is dissipated by the flow of electrons downward and positive ions upward along a complex network of paths. If you knew the effective resistance of each of the discharge paths shown above, and the potential difference between the cloud and the ground, how would you calculate the current in each path? In this unit you will learn how to use the principles of charge and energy conservation to develop a method for determining currents for each path in a complex circuit.

UNIT 23: CIRCUIT ANALYSIS

OBJECTIVES

1. To learn to apply the concept of potential difference, or voltage, to explain the action of a battery and understand the distribution of voltages in both series and parallel circuits.

2. To understand and apply the relationship between potential difference and current for a resistor (Ohm's law).

3. To find a mathematical description of the flow of electric current through different elements in more complex circuits (Kirchhoff's laws).

23.1 OVERVIEW

In the last unit we saw that in a series circuit, the current is the same through all the elements. In a parallel circuit, the current divides among the different branches (with the current through the battery equaling the sum of the currents in each branch). We found that making a change in one branch of a parallel circuit does not affect the current flowing through other parallel branches, but that changing one part of a series circuit affects all elements in series. In other words, how light bulbs are arranged, or connected together, makes a big difference in the behavior of the circuit, including whether multiple bulbs offer a larger or smaller resistance to current flow.

In this unit we first examine the role of the battery in causing a current to flow and quantitatively measure potential differences across different parts of series and parallel circuits. We then explore the concept of resistivity and the relationship between the current through a resistor and the voltage across the resistor. In addition, we'll measure the effective resistance of multiple resistors when they are wired in series and in parallel. Finally, we will formulate the rules for calculating the current in different parts of a complex circuit consisting of many resistors and/or batteries wired in series and parallel.

POTENTIAL DIFFERENCES IN CIRCUITS

23.2 POTENTIAL DIFFERENCE WHEN BATTERIES ARE COMBINED

In the following activities, we will explore voltages and currents in circuits when batteries are combined. To do this, you will need the following items:

- 3 D-cell batteries, 1.5 V
- 3 D-cell holders
- 3 #14 bulbs (with identical brightness on own)
- 3 #14 bulb holders
- 6 alligator clip leads, > 10 cm
- 2 SPST switches
- 1 voltmeter
- 1 ammeter

We have already seen what happens to the brightness of the bulb in Fig. 23.1(a) when a second bulb is added in series, as in Fig. 23.1(b). The two bulbs in series are dimmer than the original bulb because their combined resistance to current is larger, which results in less current flow through the bulbs.

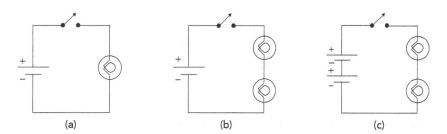

Fig. 23.1. Series circuits with (a) one battery and one bulb; (b) one battery and two bulbs; and (c) two batteries and two bulbs.

Let's make some predictions about the result of adding a second bulb *and* a second battery to the "standard" circuit of Fig. 23.1(a).

23.2.1. Activity: Adding a Second Battery and Bulb in Series

a. Predict what will happen to the brightnesses of the bulbs if you connect a second battery in series with the first and at the same time you add a second bulb in series, that is, going from Fig. 23.1(a) to Fig. 23.1(c)?

b. Build the circuits shown in Fig. 23.1(a) and Fig. 23.1(c). Briefly close both switches and compare the brightness of the bulbs. Did your observation agree with your prediction? Briefly explain.

c. What happens to the resistance of a circuit as more bulbs are added in series? Explain why you think the current appeared to stay the same in Fig. 23.1(c), even though a second bulb was added in series.

Next, you'll compare the brightness of the bulb in the circuit of Fig. 23.2 to the brightness of the single bulb in the standard circuit.

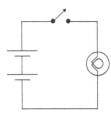

Fig. 23.2. Series circuit with two batteries and one bulb.

23.2.2. Activity: Adding Only a Second Battery

a. What do you predict will happen to the brightness of a single bulb if a second battery is added in series? Explain the reasons for your prediction.

b. Build the circuit in Fig. 23.2. **Note**: *Only close the switch for a brief moment to observe the brightness of the bulb; otherwise, you might burn out the bulb.* Compare the brightness of the bulb in Fig. 23.2 to that of the bulb in Fig. 23.1(a). Did your observation agree with your prediction?

c. How does increasing the number of batteries in series appear to affect the current in the circuit?

We just saw that adding two batteries in series increases the amount of the current in the circuit (assuming the rest of the circuit remains unchanged). Let's explore potential differences more carefully as batteries are wired in series and parallel to see if we can develop rules to describe this behavior. Figure 23.3 shows a single battery, two batteries connected in series, and two batteries connected in parallel (all batteries are identical).

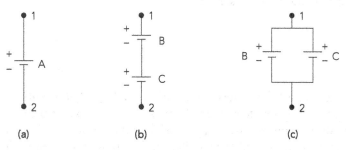

(a)   (b)   (c)

Fig. 23.3. (a) a single battery, (b) two batteries connected in series, and (c) two batteries connected in parallel.

23.2.3. Activity: Batteries in Series and Parallel

a. Assume the potential difference (voltage) between points 1 and 2 in Fig. 23.3(a) is measured to be 1.5 V. Predict the potential difference between points 1 and 2 in the *series connection* shown in Fig. 23.3(b). Similarly, predict the potential difference between points 1 and 2 in the *parallel connection* shown in Fig. 23.3(c). Briefly explain your reasoning.

b. Using three separate batteries, measure the voltages of batteries A, B, and C (measure the *individual* voltages, or when they are disconnected from each other). Record the voltage measured for each battery below.

Potential Difference Across Individual Batteries:

Battery A: _____ Volts

Battery B: _____ Volts

Battery C: _____ Volts

c. If you look carefully at the side of the battery, it should indicate that the expected potential difference across the two terminals is nominally 1.5 V. How do your measured values agree with those marked on the batteries? (If one is significantly lower, you should swap it for a fresh one.)

d. Now connect the batteries in series as in Fig. 23.3(b). Measure the potential difference between points 1 and 2 and record your value below.

Potential difference across B and C in series: _____ Volts

e. Does your measured value agree with your prediction? If not, think again about your reasoning.

f. Now connect the batteries in parallel as in Fig. 23.3(c). **Note**: You probably won't be able to connect them exactly as shown in the diagram; just be sure to connect them in parallel, and then touch one lead of the voltmeter to one side of the parallel combination and the other lead of the voltmeter to the other side. Measure the potential difference between points 1 and 2 and record your value below.

Potential difference across B and C in parallel: _____ Volts

g. Does your measured value agree with your prediction? If not, think again about your reasoning. **Hint**: This can be a bit tricky; you may find it helpful to think about the gravitational analogy for potential difference.

h. The voltage of two identical batteries in parallel was the same as the individual batteries. Why might someone want to wire two batteries in parallel? What purpose might this serve in a circuit?

You should have found that the voltages of batteries in series add, while voltage of batteries in parallel is the same as an individual battery[1]:

$$\Delta V_{\text{series}} = \Delta V_1 + \Delta V_2 + \cdots \quad (\text{batteries in series}) \qquad (23.1)$$

$$\Delta V_{\text{parallel}} = \Delta V_1 = \Delta V_2 = \cdots \quad (\text{batteries in parallel}) \qquad (23.2)$$

23.3 POTENTIAL DIFFERENCES IN SERIES AND PARALLEL CIRCUITS

In Unit 22 we examined several simple series and parallel circuits and discussed the relative brightness of different bulbs using arguments related to the resistance to current flow. In this section we return to some of those same circuits and quantify voltage and current measurements in order better understand how circuits operate. This will also set us up to use Ohm's law to answer questions that cannot easily be determined by current arguments alone.

To do the activities in this section, you will need:

- 2 D-cell batteries, 1.5 V
- 2 D-cell holders
- 2 #14 bulbs (with identical brightness on own)
- 2 #14 bulb holders
- 6 alligator clip leads, > 10 cm
- 1 SPST switch
- 1 voltmeter
- 1 ammeter

Let's begin with the circuit containing two bulbs in series with a single battery. Figure 23.4 shows this circuit, along with a number of points in the circuit that we will use as a reference for measuring potential difference. For example, the potential difference across bulb B involves measuring the potential difference between the points 3 and 4. We will indicate the voltage of point 3 with respect to point 4 as ΔV_{34}.[2] To measure ΔV_{34} experimentally, the red lead of the voltmeter should be connected to point 3, while the black lead should be connected to point 4.

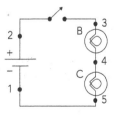

Fig. 23.4. A series circuit with one battery and two bulbs.

---

[1] You should not attempt to connect two batteries with different voltages in parallel, as it can damage the batteries.

[2] Since ΔV_{34} tells one how much higher the potential is at point 3 as compared to point 4, ΔV_{34} will be positive if point 3 is at a higher potential than point 4 (or negative if point 3 is at a lower potential than point 4).

23.3.1. Activity: Voltages in Series Circuits

a. Assuming bulbs B and C in Fig. 23.4 are identical, how do you predict the potential difference across bulb B (ΔV_{34}) will compare to the potential difference across the battery (ΔV_{21})? Similarly, what about the potential difference across bulb C (ΔV_{45})?

b. Predict how the potential difference across the series combination of bulbs B and C (ΔV_{35}) will compare to the potential difference across the battery (ΔV_{21}).

c. Wire up the circuit and test your predictions by measuring the voltages and recording their values. **Note**: Be sure the switch is *closed* when you make your measurements. Did your observations agree with your predictions?

Potential difference across the battery (ΔV_{21}): _____ Volts

Potential difference across bulb B (ΔV_{34}): _____ Volts

Potential difference across bulb C (ΔV_{45}): _____ Volts

Potential difference across bulbs B and C in series (ΔV_{35}): _____ Volts

d. Based on your observations, formulate a rule for how voltages across individual bulbs *in series* combine to give the total voltage across the entire series combination.

You should have found that total voltage across the series combination splits up equally among the bulbs (assuming they are identical). Mathematically, in a series combination we have

$$\Delta V_{\text{single bulb}} = \frac{\Delta V_{\text{total}}}{\#\text{bulbs}} \quad (\text{bulbs in series}) \qquad (23.3)$$

which implies that *voltages add in series*. Note that this is just like how batteries combine in series—their voltages add.

Let's now consider a circuit with two bulbs in parallel, as shown in Fig. 23.5. We will include a switch in this circuit in such a way that we can look at a single bulb as well. We also will use two batteries in parallel to avoid draining the energy from a single battery.

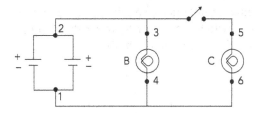

Fig. 23.5. Parallel circuit with two bulbs and a parallel battery setup.

23.3.2. Activity: Voltages in Parallel Circuits

a. Predict what will happen to the potential difference across the batteries (ΔV_{21}) when you close the switch. Will it change or remain (essentially) the same?

b. Predict how the potential difference across bulb B (ΔV_{34}) will compare to the voltage of the batteries (ΔV_{21}) for both the case of the switch open and the switch closed.

c. Predict how the potential difference across bulb C (ΔV_{56}) will compare to the voltage of the battery (ΔV_{21}) for both the case of the switch open and the switch closed.

d. Now wire up the circuit and measure the potential difference across the batteries (ΔV_{21}), across bulb B (ΔV_{34}), and across bulb C (ΔV_{56}) when the switch is open and when the switch is closed. Record your results to one decimal place (tenths of a volt) in the table below.

	Voltage across battery	Voltage across bulb B	Voltage across bulb C
Switch open			
Switch closed			

   e. Do your measurements agree with your predictions? Did closing and opening the switch *significantly* affect the voltage across the battery? What about the voltage across bulb B?

   f. Did closing and opening the switch significantly affect the voltage across bulb C? Under what circumstances is there a potential difference across a bulb?

   g. Based on your observations, formulate a rule for how voltages across individual bulbs *in parallel* compare to each other (and to the battery to which they are connected).

   h. Although we didn't measure it, based on what we did in Unit 22, you should be able to determine how the total current flowing through the battery changes when you close and open the switch. How does the total current through the battery when the switch is closed compare to the total current when the switch is open? Briefly explain.

   i. Based on your results, is a battery a constant current source, a constant voltage source, or neither? (We already considered this in Unit 22, but it is worth reemphasizing!)

You should have found that the voltage across each bulb in parallel is the same (and equal to the battery to which they are connected). Mathematically, in a parallel combination we have

$$\Delta V_{\text{parallel}} = \Delta V_1 = \Delta V_2 = \cdots \ \left(\text{bulbs in parallel}\right) \qquad (23.4)$$

Note this is just like how batteries combine in parallel. While the voltages are the same across elements in parallel, the currents can be different.

Summarizing what we have seen thus far, we can say the following:

Elements in series: currents are the same, voltage "divides"
- voltages across all elements add up to the total voltage
- voltages across identical elements are the same

Elements in parallel: voltages are the same, current "divides"
- currents through all elements add up to the total current
- currents through identical elements are the same

How exactly the voltages and currents divide will be determined during the rest of this unit.

23.4 OHM'S LAW: QUANTITATIVELY RELATING VOLTAGE AND CURRENT

We have seen that there is only a potential difference across a bulb when there is current flowing through it. We also have said that a larger resistance to current flow implies a smaller amount of current. In the next activity we will quantify how the current through an element like a light bulb depends on the potential difference across it. However, instead of a light bulb we'll use a carbon *resistor*. Resistors are designed to resist the flow of current (unsurprisingly) and have essentially the same resistance to current no matter how much current is flowing. The circuit element for a resistor is a jagged line, as shown in Fig. 23.6.

You will need the following equipment:

R

Fig. 23.6. Circuit diagram resistor symbol.

- 3 D-cell batteries, 1.5 V
- 3 D-cell holders
- 1 resistor, approximately 75Ω
- 1 SPST switch
- 1 voltmeter
- 1 ammeter

23.4.1. Activity: Relationship Between Current and Voltage

a. How do you predict the potential difference across a resistor will be related to the current through the resistor. Talk it over with your group and sketch a graph of your predicted relationship of voltage vs current, using ΔV to indicate the potential difference (voltage) and i to represent the current.

b. Set up a circuit to test your prediction by connecting the resistor to a different number of batteries in series. Be sure that the voltmeter is measuring the voltage *across* the resistor, and that the ammeter is measuring the current flowing *through* the resistor. Make a sketch of your circuit diagram and summarize your data in the table. We have gotten you started with the first data point: no batteries implies no voltage and no current!

Number of batteries	ΔV (Volts)	i (Amps)
0	0.0	0.0
1		
2		
3		

c. Use graphing software to make a plot of ΔV vs i and perform a linear fit on the plot.[3] Sketch or affix a printout of your plot, including the result of your fit. Do your results agree with your prediction?

Resistivity and Resistance

You should have found a fit showing that the potential difference across the resistor is roughly *proportional* to the current through the resistor. The slope of the graph defines the resistance to current flow for this circuit element: a larger slope implies that a larger potential difference is required to produce the same current. The resistance to current flow depends on both the material the resistor is made of, as well as the physical geometry of the resistor. A related concept that does not depend on the geometry of the object is the *resistivity*.

The resistivity of a material is a measure of how strongly the material resists the flow of current. Resistivity is intrinsic to the type of material and depends on the microscopic details of the electron configuration and lattice structure of the material. We won't go into the details of how one determines the resistivity, but simply say that it's related to the number of charge carriers and how quickly they can move through the material (the drift speed).

Resistivity is denoted by the Greek letter "rho" (ρ) and has units of ohm-meters, where one ohm (the unit of resistance) is defined to be one volt per amp and is represented with the Greek letter "Omega" (Ω): $1\ \Omega \equiv 1\ \text{V}/1\ \text{A}$. Resistivity values for standard materials can range over many orders of magnitude. For example, copper is a very good conductor (low resistivity), having a

---

[3] Since ΔV is the independent variable in our experiment, it would be more common to plot i vs ΔV. However, the expected relationship is such that it makes more sense to plot ΔV vs i in this case. Depending on which software you are using to plot the data, you may need to adjust your column order to ensure you are plotting ΔV vs i.

value of $\rho \approx 2 \times 10^{-8}$ Ω m at room temperature, while carbon is not a great conductor (higher resistivity) and has a value of $\rho \approx 10^{-4}$ Ω m in its graphite form. On the other hand, a good insulator such as rubber will have a resistivity of $\rho \approx 10^{+13}$ Ω m.

The *resistance* of a circuit element depends on both the resistivity of the material, as well as the dimensions (the geometry) of the circuit element. While resistivity is intrinsic to the type of material, the resistance is what matters when putting an element into a circuit. The relationship between resistance R and resistivity ρ is given (without proof) by:

$$R = \rho \frac{L}{A} \tag{23.5}$$

where L is the length of the object, A its cross-sectional area (see Fig. 23.7), and R has units of ohms.

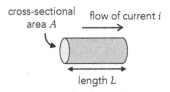

Fig. 23.7. The resistance R of an element depends on the resistivity ρ of the material as well as the cross-sectional area A and length L of the element. Current is assumed to flow along the length of the element.

You can probably convince yourself of the (linear) dependence of resistance on length—even a material with low resistivity can provide a large resistance to current if its length is really long! The (inverse) dependence of resistance on cross-sectional area of the material may not be as obvious. The larger the cross-sectional area, the more "parallel paths" there are for the charges to take. This means that more current can flow for the same "pushing force" from the battery.

Experimentally, it is the resistance that matters in a circuit, and the resistance of a resistor can be determined directly from your data in Activity 23.4.1. For "ohmic" materials such as resistors, the resistance is simply the slope of the ΔV vs i graph.[4] Mathematically, we can write this as

$$R = \frac{\Delta V}{i} \tag{23.6}$$

where R is the resistance of a circuit element, ΔV is the potential difference across the element, and i is the current flowing through the element.

---

[4] A circuit element (like a light bulb), whose resistance changes with the amount of current, is called *non-ohmic*. Although the ΔV vs i graph for a non-ohmic element will not have a constant slope, one can still think about the instantaneous slope of ΔV vs i at any given point to get an effective resistance at that current value.

23.4.2. Activity: Statement of Ohm's Law

a. Use your fit from Activity 23.4.1 to determine the experimentally-measured value of R for your resistor.

b. Rearrange the mathematical relationship between resistance, voltage, and current given in Eq. (23.6) to solve for ΔV on one side (this is straightforward—don't overanalyze!). It is more common to see Eq. (23.6) in this form.

You have experimentally determined that for ohmic circuit elements, the potential difference across the element is proportional to the current through the element. This is referred to as *Ohm's law*, typically written as

$$\Delta V = iR \quad \left(\text{Ohm's Law}\right) \tag{23.7}$$

The Sign in Ohm's Law

Equation (23.7) gives the potential difference in terms of the current through a resistor and its resistance. However, Eq. (23.7) on its own does not tell us whether the potential difference should be positive or negative. People usually talk about a positive voltage when discussing the potential difference across a resistor, but as we'll see, the voltage change in Ohm's law is actually negative when moving in the direction of current flow. This sign comes from two facts: (1) the positive terminal of the battery is at a higher voltage than the negative terminal, and (2) the convention that (positive) current flows out of the positive terminal of the battery and into the negative terminal.

Figure 23.8 shows the basic idea, where a current i flows clockwise around the loop. The resistor has a potential difference across its two ends, and from Fig. 23.8, we can see that the top of the resistor (connected to the positive terminal of the battery), must be at a higher potential than the bottom of the resistor (which is connected to the negative terminal of the battery) so that

$$\Delta V_{34} = V_3 - V_4 = V_2 - V_1 = \Delta V_{21} > 0$$

This expression confirms that the sign of the change in voltage while passing through a resistor in the direction of current flow (from point 3 to point 4 in Fig. 23.8) is *negative*:

$$\Delta V_{43} = V_4 - V_3 < 0 \quad \left(\text{when moving in the direciton of current}\right)$$

We will return to this fact later in the unit.

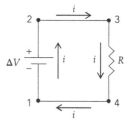

Fig. 23.8. A single-loop circuit consisting of a battery and resistor.

RESISTORS AND RESISTANCE NETWORKS

23.5 RESISTORS AND RESISTANCE MEASUREMENT

If one wants a constant resistance in a circuit, it is customary to use a carbon resistor instead of a light bulb. Carbon resistors are inexpensive to manufacture, tend not to "burn out," and can be produced with just about any value of resistance. More importantly, unlike a bulb, the resistance of a carbon resistor is (more-or-less) independent of the amount of current flowing through it.

A typical carbon resistor contains a form of carbon known as graphite suspended in a hard glue binder, and is usually surrounded by a plastic case with a color code painted on it (see Fig. 23.9). The colored bands signify its resistance value in ohms, and each color has a number associated with it, as shown in the following table.

A table representing the resistor code	
Black = 0	Blue = 6
Brown = 1	Violet = 7
Red = 2	Gray = 8
Orange = 3	White = 9
Yellow = 4	Silver = ±10%
Green = 5	Gold = ±5%

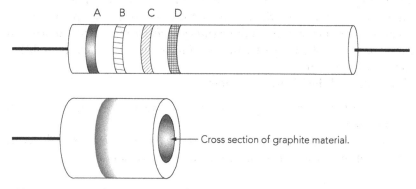

Fig. 23.9. Top: A carbon resistor with color bands showing the value of resistance (and tolerance). Bottom: A cutaway view of a carbon resistor showing the cross-sectional area of the graphite material.

The value in ohms is given using the following formula

$$R = AB \times 10^C \pm D$$

where AB means the A digit is placed beside the B digit (not A times B) and represents a single number. The colors on bands A, B, and C represent the digits shown in the table above, and the silver or gold D band represents the "tolerance" of the resistor (the guaranteed precision of the resistance value). If there is no D band, the tolerance is ±20%.

The resistor code formula is probably best understood with an example. Consider a resistor with color bands of Yellow-Gray-Red-Silver. Looking at the chart above to covert colors to numbers, we see this resistor has $A = 4$, $B = 8$, $C = 2$, and $D = \pm 10\%$. Plugging these values into our resistor code formula gives

$$R = 48 \times 10^2 \pm 10\% = 4800 \pm 480 \ \Omega$$

This resistor has a specified value of $4800 \ \Omega = 4.8 \ k\Omega$ and is guaranteed by the manufacturer to be within the range 4320–5280 Ω.

A multimeter can also function as an *ohmmeter* and experimentally measure the resistance of a resistor. To do so, turn the dial on the meter to the section marked resistance (or ohms, or Ω). There may be different settings in this section; the numbers shown correspond to the *maximum* resistance you can measure on each setting (with "k" representing kilo or 10^3, and "M" representing mega or 10^6). Connect the black multimeter probe to the common input and the red probe to the input designated for measuring resistance. The meter shows the measured resistance between the two probes.

If the probes are not connected to anything, your meter will probably show "OL," which indicates "overload." With nothing but air between the two probes, there is essentially infinite resistance, and the meter tells you this by saying overload. To measure the resistance value of a resistor, touch the red probe to one side of the resistor and the black probe to the other side. You need to make good contact between the metal of the probe and the metal of the resistor—for resistances much less than a megaohm, it is okay to "pinch" the two together with your finger, assuming you make good metal-to-metal contact.

In the first activity, we will determine the rated resistance in ohms of a few different resistors and then check the results by measuring the resistance using a multimeter. In the following activities, we'll measure the resistance of different combinations of resistors.

You will need the following equipment:

- 5 assorted color-coded carbon resistors, two of which have identical resistance
- 4 alligator clip leads *with low-resistance* (full-copper test leads are best)
- 1 multimeter with resistance measurement capability

23.5.1. Activity: Decoding and Measuring Resistors

a. Decode three to five resistors using the color bands and the resistor code formula. Then measure the resistance of each resistor using the multimeter. Determine the percent discrepancy between the coded value of resistance and the value you determine experimentally. Use this to determine whether the measured value is within the rated tolerance. Record all information in the following table. In general, the percent discrepancy is given by the expression

$$\text{Percent Discrepany} = \left| \frac{\text{Accepted Value} - \text{Measured Value}}{\text{Accepted Value}} \right| \times 100\%$$

Color sequence	Coded R (Ω)	Measured R (Ω)	Calculated percent discrepancy	Percent tolerance	Within rated tolerance?
1					
2					
3					
4					
5					

b. How do you think the ohmmeter functionality works? In other words, speculate as to how the multimeter manages to determine the resistance of a circuit element. **Hint**: Remember that you have already used the multimeter to determine both voltage and current.

Resistance Networks: Resistors in Series and Parallel

Equation (23.5) says that the resistance of an element is directly proportional to its length and inversely proportional to its cross-sectional area. Thus, it is possible to control the resistance value simply by changing the length and/or cross-sectional area of the carbon core of a resistor. We can use this idea to provide a conceptual understanding of how resistors combine when connected in series or parallel. Figure 23.10 shows two resistors wired in series and parallel. The total resistance of an arrangement of resistors is known as the *equivalent resistance* R_{eq} of the combination. In the next activity, we'll consider how resistors act when wired *in series*.

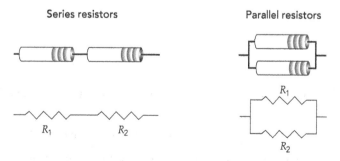

Fig. 23.10. Carbon resistors wired in series (left) and in parallel (right).

23.5.2. Activity: Equivalent Resistance for Series Wiring

a. Suppose you have three carbon resistors. What do you think the equivalent resistance will be if the resistors are wired in *series*? Explain the reasons for your prediction based on your previous observations with bulbs, as well as Eq. (23.5).

b. Pick out three resistors and use the multimeter to measure the resistance value of each resistor *individually*:

$$R_1 = \underline{\hspace{2cm}} \Omega$$

$$R_2 = \underline{\hspace{2cm}} \Omega$$

$$R_3 = \underline{\hspace{2cm}} \Omega$$

c. Using these numbers, determine your predicted resistance when these resistors are connected in series:

$$\text{Predicted } R_{eq} = \underline{\hspace{2.5cm}} \Omega$$

d. Now connect the three resistors in series and measure the actual resistance of the series network. You can use alligator leads to clip the resistors together, "pinch" the resistor ends together, or "twist" the resistor ends together; the important point is to have good metal-to-metal contact at each junction.

$$\text{Measured } R_{eq} = \underline{\hspace{2.5cm}} \Omega$$

e. How does the measured value compare to your prediction? Devise a general rule that describes the equivalent resistance when n total resistors are wired in series. Use the notation R_{eq} for the equivalent resistance and $R_1, R_2, \ldots, R_n$ to represent the values of the individual resistors.

You should have found that when resistors are connected in series, their resistances *add*:

$$R_{eq} = R_1 + R_2 + R_3 + \cdots + R_n \quad (\text{resistors in series}) \qquad (23.8)$$

For the case of identical resistors, it's pretty easy to see how this comes about physically: two identical resistors in series can be thought of as a single resistor with twice the length (but the same cross-sectional area). Therefore, the series combination of two resistors should offer twice the resistance. In the next activity, we'll consider how resistors act when connected *in parallel*.

23.5.3. Activity: Equivalent Resistance for Parallel Wiring

a. Suppose you have *two identical* carbon resistors (they don't have to be identical, but that is probably the easiest way to start). What do you think the equivalent resistance to the flow of current will be if the resistors are wired in *parallel*? Explain the reasons for your prediction based on your previous observations with bulbs, as well as Eq. (23.5).

b. Pick out two carbon resistors with an *identical color code* and use the multimeter to measure the resistance value of each resistor *individually*:

$$R_1 = \underline{\hspace{2cm}} \Omega$$

$$R_2 = \underline{\hspace{2cm}} \Omega$$

c. Using these numbers, determine your predicted resistance when these resistors are connected in parallel:

$$\text{Predicted } R_{eq} = \underline{\hspace{2cm}} \Omega$$

d. Now connect the resistors in parallel and measure the actual resistance of the parallel network. For parallel wiring, it can sometimes be challenging to figure out where to connect the multimeter probes. Remember, you want the red probe connected to one "side" of the parallel combination and the black lead connected to the other "side."

$$\text{Measured } R_{eq} = \underline{\hspace{2cm}} \Omega$$

e. How does the measured value compare with what you predicted for the case of two, identical resistors?

f. Next pick out *three different* resistors. In the space below, draw a diagram for these three resistors wired in *parallel* and label the diagram with the values R_1, R_2, and R_3. Hook up the three resistors in parallel and measure the value of the equivalent resistance of the network.

$$\text{Measured } R_{eq} = \underline{\hspace{2cm}} \Omega$$

g. The equation to calculate the equivalent resistance in parallel is not as easy to come up with as the one for resistors in series. It turns out that resistors in parallel obey

$$\frac{1}{R_{eq}} = \frac{1}{R_1} + \frac{1}{R_2} + \frac{1}{R_3} + \cdots + \frac{1}{R_n} \quad \text{(resistors in parallel)} \quad (23.9)$$

Below, use this formula to calculate the expected value of R_{eq} for both cases we examined. **Note**: When using your calculator, it can be easy to forget to "invert" at the end to arrive at R_{eq} (as opposed to $1/R_{eq}$)!

1. For two identical resistors wired in parallel:

    Calculated R_{eq} = _____Ω

    Measured R_{eq} = _____Ω

2. For three different resistors wired in parallel:

    Calculated R_{eq} = _____Ω

    Measured R_{eq} = _____Ω

h. For the case of *two* resistors, one can simplify the formula to a form that is a little easier to use. Starting with Eq. (23.9) for the case of two resistors, *mathematically prove* that you can rewrite the equivalent resistance as[5]

$$R_{eq} = \frac{R_1 R_2}{(R_1 + R_2)} \quad \text{(two resistors in parallel)} \quad (23.10)$$

23.6 RESISTANCE NETWORKS

Now that we know the basic equations to calculate equivalent resistance for series and parallel combinations we can tackle the question of how to find the equivalent resistance for a complex network of resistors. The approach is to calculate the equivalent resistance for a subsection of the complex network, and then use that result in the calculation for the next subsection.

One typically wants to start as deep into the circuit as possible where, for example, you see a subset of resistors that are simply in series (or parallel) with each other. In this case, Eq. (23.8) or (23.9) allows you to determine the equivalent resistance of the subset. One can then simplify this subsection by redrawing the circuit with a new *equivalent resistor* that represents the equivalent resistance of that combination. That equivalent resistor acts just like a single resistor in all subsequent calculations.

For example, consider the resistor network shown in Fig. 23.11. The original network is shown in the top panel, where there are four total resistors comprised of two types with values R_1 and R_2.

---

[5] For the case of two identical resistors in parallel, one can think of the parallel combination as a single resistor with the same length as the individual resistors but twice the cross-sectional area. According to Eq. (23.5), we would expect $R_{eq} = R/2$ for two identical resistors in parallel, consistent with Eqs. (23.9) and (23.10).

Step 1: We can immediately spot that the R_1 and R_2 resistors along the top parallel branch are in series with each other. This allows us to define a new equivalent resistance we call R_3, where $R_3 = R_1 + R_2$. The middle panel of Fig. 23.11 shows the circuit redrawn using this new equivalent resistor R_3 to represent the two resistors we have replaced.

Step 2: Looking at the redrawn circuit, we can now see that on the left side, R_3 is in parallel with R_2. We use our parallel resistor formula to replace these two resistors with a new, equivalent resistor we call R_4. Once again, we redraw the circuit, putting the equivalent resistor R_4 in place of R_2 and R_3 (see bottom panel of Fig. 23.11).

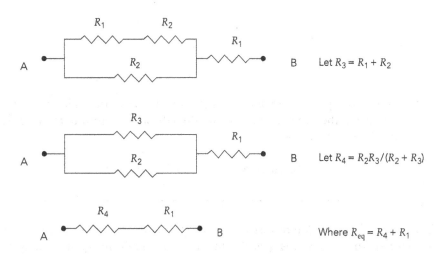

Fig. 23.11. A sample resistor network with simplifications.

Step 3: Finally, in the bottom panel equivalent resistor R_4 is in series with R_1, so we can combine these two in series. The final equivalent resistance is then simply $R_{eq} = R_4 + R_1$.

In order to complete the equivalent resistance activities, you will need the following equipment:

- 3 carbon resistors, 100 Ω
- 3 carbon resistors, 220 Ω
- 6 alligator clip leads *with low-resistance* (full-copper test leads are best)
- 1 multimeter

23.6.1. Activity: The Equivalent Resistance for a Network

a. Figure 23.12 shows a network of resistors that involves two different resistance values (R_1 and R_2). Choosing $R_1 = 100\ \Omega$ and $R_2 = 220\ \Omega$, calculate the equivalent resistance between points A and B for the network.[6] Remember to begin as deep in the circuit as possible, where it is clear which resistors are in series or parallel. You should show your calculations on a step-by-step basis, indicating which resistors you are combining during each step.

---

[6] If you have different value resistors than those specified, use the smaller value for R_1 and the larger value for R_2.

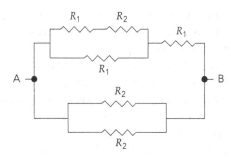

Fig. 23.12. A complex network of resistors.

b. Now wire up this network of resistors and check your calculation by measuring the equivalent resistance directly with a multimeter. Does it match your calculation?

$$\text{Measured } R_{eq} = \underline{\hspace{3cm}}\ \Omega$$

We now know how to combine resistors in complex networks to form equivalent resistance(s). To see how this is useful in an actual circuit, in the next activity we use the concept of equivalent resistance to calculate voltages and currents in a circuit.

23.6.2. Activity: Equivalent Circuits

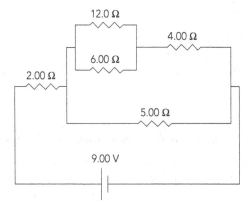

Fig. 23.13. A network of resistors connected to a battery.

a. Consider the circuit shown in Fig. 23.13. Find the equivalent resistance of the network of resistors. Then, redraw the final circuit with the battery and a single, equivalent resistor representing the equivalent resistance.

b. Although the equivalent resistor you drew in the simplified circuit is not real (there is no actual resistor with this exact value in the circuit), we can think of it as if it were real to help us understand how the circuit behaves. For example, use the equivalent resistance of the circuit to determine the total current coming from the battery.

c. How much of this current through the battery flows through the 2 Ω resistor? Use this information and Ohm's law to determine the potential difference (the "voltage drop") across the 2 Ω resistor.

d. Based on your answer to part (c), what is the potential difference across the *remaining* combination of resistors?

e. Based on your answer to part (d), what is the current through the 5 Ω resistor?

Hopefully, it's clear that the concept of equivalent resistance is a powerful tool for analyzing circuits. Although equivalent resistors do not actually exist, we can use the idea of them to calculate currents and voltages as you did in Activity 23.6.2. In fact, one could continue using equivalent resistances and Ohm's law to determine the voltage across, and current through, every resistor in Fig. 23.13. However, there are some circuits where blindly applying Ohm's law will simply not work, so we need another approach. But before we present how to do the full analysis of a complex circuit, there is one more concept we need to introduce.

23.7   POWER IN CIRCUITS

In Unit 22 we discussed how the fact that current is not "used up" in a bulb indicates that the rate energy is dissipated in a bulb (the *power*) cannot depend only on current. After all, even though the bulb is clearly "dissipating energy," the current is same before and after the bulb. As you may have guessed at this point, the missing piece to this puzzle is that the power dissipated also depends on voltage. We are now ready to formalize this relationship.

Consider a simple circuit of a single battery and resistor connected by wires as in Fig. 23.14. The battery maintains a potential difference ΔV_b across the two sides of the resistor, which drives a current through the resistor from point 3 to point 4. The instantaneous current is $i = dq/dt$, where dq is the (infinitesimal) amount of charge that flows past a given point in time dt. As an amount of charge dq moves from one side of the resistor to the other, it experiences a change in voltage $\Delta V_{43} = -\Delta V_b$, where the negative sign is due to the fact that voltage *decreases* as the (positive) current moves from point 3 to point 4. Using the relationship between electrical potential energy and electric potential in Eq. (21.10), the change in potential energy is found to be

$$dU = dq\left(-\Delta V_b\right) = -i\, dt\left(\Delta V_b\right)$$

This change in potential energy corresponds to an amount of work (see Eq. (21.9)) $dW = -dU = +i\, dt(\Delta V_b)$. Finally, we can determine the instantaneous power supplied to the resistor from Eq. (10.6):

$$P = \frac{dW}{dt} = i\,\Delta V_R \quad \text{(Electrical Power)} \tag{23.11}$$

where we have used ΔV_R to signify the (positive) voltage across the resistor (in general, this will not be the same as the battery voltage). In words, the instantaneous (electrical) power supplied to a resistor is the product of the current through the resistor and the change in voltage across the resistor. **Note:** All quantities in Eq. (23.11) are assumed to be positive. Because power is a rate of energy transfer, we can think of the battery as transferring energy to the resistor in Fig. 23.14 at a rate given by $i\,\Delta V_R$ (you may also hear it said that the resistor *dissipates* energy at a rate $i\,\Delta V_R$). As usual, the unit of power is the watt, which from Eq. (23.11) is equivalent to an ampere-volt.

While Eq. (23.11) is the general result, it can be expressed in other ways when dealing with ohmic devices such as resistors. The next activity looks at these relationships.

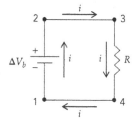

Fig. 23.14. A single-loop circuit with a battery and resistor. Current flows clockwise from the positive terminal of the battery, down through the resistor, and back through the battery.

23.7.1. Activity: Power in Ohmic Devices

a. Use the result for electrical power in Eq. (23.11), along with Ohm's law in Eq. (23.7) (valid for ohmic devices like resistors), to write the power in terms of the current and resistance (no potential difference).

b. Do something similar, this time writing the power in terms of the potential difference and resistance (no current).

KIRCHHOFF'S CIRCUIT RULES AND ANALYSIS

23.8 KIRCHHOFF'S RULES

Suppose we wish to calculate the currents in various branches of a circuit that has many components wired together in a complex array. In such circuits, simplification using series and parallel combinations can be challenging, if not impossible. Instead, we state and apply a formal set of rules known as Kirchhoff's Circuit Laws (or just Kirchhoff's Laws, or Kirchhoff's Rules) to use in the analysis. These rules are not really new to us, as we have touched on them before. However, we have not stated them formally until this point.

Kirchhoff's Rules

1. *Current Junction (or Node) Rule*: A junction, or node, is where two or more wires join/split. The Junction Rule says that the sum of all the currents entering a junction must equal the sum of all currents leaving the junction. In other words, current into the junction equals current out of the junction. This rule is based on charge conservation and can be thought of similarly to water in pipes (or hoses). Mathematically, this rule is written as

$$i_{\text{into junction}} = i_{\text{out of junction}} \qquad (23.12)$$

2. *Voltage Loop Rule*: We know that current must have a closed circuit to flow, meaning that there will be (at least) one "loop" in every circuit where charge carriers can follow a complete, closed path. The Loop Rule says that around *any* closed loop in a circuit, the sum of all voltage differences provided by batteries and occurring across resistors (or other circuit elements) must equal zero. This rule is based on energy conservation and can be thought of similarly to moving a marble in a complete loop in our pegboard model ("up" due to the action of the battery and then "down" the pegboard back to where it started). Mathematically, this rule is written as

$$\Delta V_{\text{closed loop}} = 0 \qquad (23.13)$$

Kirchhoff's Rules are always true and can be applied to any circuit. But for most simple circuits they are not necessary, and the concepts of equivalent resistance and Ohm's law are enough to fully analyze the circuit. However, there are some types of circuits where Kirchhoff's rules are necessary, and these tend to be characterized by having *multiple batteries in different loops of the circuit*. Nevertheless, we will start with a simple circuit to demonstrate the approach.

Consider the single-loop circuit shown in Fig. 23.15. We are interested in finding the total current in the circuit. In the next activity you will follow the steps for applying Kirchhoff's rules to solve for the current, and then compare this result to what you get using equivalent resistance and Ohm's law.

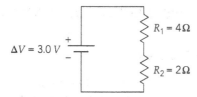

Fig. 23.15. A one-battery, two-resistor series circuit for demonstrating Kirchhoff's rules.

23.8.1. Activity: Steps for Applying Kirchhoff's Rules

Step 1: The first step in Kirchhoff's rules is to assign a current to each branch of the circuit and label them i_1, i_2, etc. For this simple circuit, there is only one branch since there are no junctions of wires; the current in this circuit is the same everywhere, so we only need a single current i_1. You then *arbitrarily* assign a direction to each current. The direction chosen for each branch doesn't matter; if you choose the "wrong" direction, the sign of the current will simply turn out to be negative, telling you it flows in the opposite direction!

   a. Choose a direction for the current and draw an arrow in the circuit labeled i_1 showing the direction.

Step 2: The second step is to draw loops for the Voltage Loop Rule. Each loop you draw needs to show the starting point and the direction (because it is a closed loop, the ending point must be the same as the starting point). In our circuit there is only one loop, so we only have to do it once.

   b. Choose a starting point and direction for your voltage loop and draw a "circular arrow" labeled "1" (for loop 1).

Step 3: The third step is to apply the Voltage Loop Rule to each loop in the circuit. To do so, consider moving around the circuit in a loop, beginning at the starting location and proceeding in the direction indicated. As you move around the loop, remember three things: (1) the voltage change along a wire is zero; (2) the voltage change across a resistor is determined by Ohm's law; and (3) the voltage change across a battery is given by the voltage rating of the battery (or the measured voltage). For both (2) and (3) we also need to consider the *sign* of the voltage change:

- For *resistors*, the sign is determined by the relative directions of the current and voltage loop at the location of the resistor. If the voltage loop is in the *same direction* as the current through the resistor, the voltage change is *negative* (a voltage drop). On the other hand, if the voltage loop is in the *opposite direction* as the current through the resistor, the voltage change is *positive* (a voltage gain).
- For *batteries*, only the direction of the voltage loop matters. If the voltage loop takes you *from the negative side of the battery to the positive side*, the voltage change is *positive* (a "step up"), but if the voltage loop takes you *from the positive side of the battery to the negative side*, the voltage change is *negative* (a "step down").

You continue around the loop until you get back to your starting location (a closed loop), keeping track of the voltage change across each element. Finally, you add up all the voltage changes around the loop and set the total equal to zero.

   c. Apply the Voltage Loop Rule to Loop 1 in your circuit. Don't forget to set the sum equal to zero!

Step 4: The fourth step is to apply the Current Junction Rule to each junction in the circuit. To do so, you look at each junction of wires in the circuit and set the sum of the currents flowing into the junction equal to the sum of the currents flowing out of the junction. The simple circuit of Fig. 23.15 doesn't have any junctions, so there is nothing for us to do here! We'll wait until the next activity to see how this rule is applied.

Step 5: The fifth and final step is to mathematically solve the equation(s) for the unknown(s). Note that the "physics" is all done; the rest is algebra. Typically, one assumes that you know the voltages of the batteries and the resistances of the resistors, meaning that you want to solve for the unknown currents (one could, of course, imagine a different scenario). This procedure results in a system of equations that need to be solved for the currents, which can be done using substitution or some other mathematical technique. For our simple example, there is only one equation, and so you can immediately solve it for the current!

   d. Solve the equation from the Voltage Loop Rule in part (c) for the unknown current i_1. **Note**: If you get a negative number, it means the current flows opposite to the direction of your i_1 arrow.

   e. Finally, consider the circuit using our previous approach. Find the equivalent resistance of the resistor combination and use this, along with Ohm's law, to determine the current that flows through the battery (and hence both resistors). How does your result here compare to your result using Kirchhoff's rules in part (d)? (They should be the same!)

The last question in the previous activity shows that we can find the current for this simple circuit without going through all the steps of Kirchhoff's rules. However, consider the circuit shown in Fig. 23.16, which has "two loops" with a battery in each loop. This type of circuit does not lend itself to equivalent resistance and Ohm's law, but one can still use Kirchhoff's rules to solve for the currents.

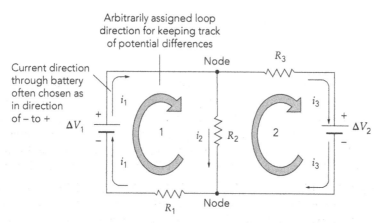

Fig. 23.16. A complex circuit in which loops 1 and 2 share the resistor R_2. The current directions and voltage loops are shown.

In the next activity we will use the steps of Kirchhoff's rules from Activity 23.8.1 to solve for the unknown currents in the circuit in terms of the problem variables. Steps 1 and 2 of Kirchhoff's rules have been done for you.

23.8.2. Activity: Kirchhoff's Rules for a More Complicated Circuit

Step 1: The currents have been labeled for you in Fig. 23.16. Note that this circuit requires three separate currents. Current i_1 is (arbitrarily) chosen to flow through resistor R_1 from right to left, and this same current must then flow up through the battery V_1. When current i_1 gets to the node (junction) at the top, it can split into two paths. Because we don't know exactly how the current splits, there must be two more unknown currents: i_2 (chosen to flow down through R_2) and i_3 (chosen to flow through R_3 to the right).

Step 2: The voltage loops have also been drawn for you. This circuit contains two separate loops, and so there are two voltage loops shown (Loop 1 on the left and Loop 2 on the right). Both loops are (arbitrarily) chosen to go clockwise, and in Step 3 we begin both loops at their respective lower-right corners.[7]

a. Apply the Voltage Loop Rule to both loops in the circuit (**Step 3**). Start each loop in the lower right corner and work your way clockwise in the direction of the loop until you get back to where you started. Follow the rules provided in Activity 23.8.1, being very careful with the signs. At the end, remember to set the sum of all voltage changes equal to zero

---

[7] There is a third possible loop for this circuit—do you see it? You could have chosen a loop all the way around the outside of the circuit. It turns out that this third loop does not provide any new information, and so we don't need it!

(separately for each loop). You should have two equations at the end, both in terms of the problem variables.

b. Apply the Current Junction Rule to both wire junctions (nodes) in the circuit (**Step 4**). Set the sum of the currents flowing into the junction equal to the sum of the currents flowing out of the junction: $i_{in} = i_{out}$. Look at the two equations you formed—what do you notice?[8]

You should have found the following set of equations:

$$\text{Loop 1: } -i_1 R_1 + \Delta V_1 - i_2 R_2 = 0 \qquad (23.14)$$

$$\text{Loop 2: } +i_2 R_2 - i_3 R_3 - \Delta V_2 = 0 \qquad (23.15)$$

$$\text{Node 1: } i_1 = i_2 + i_3 \qquad (23.16)$$

c. Assume that we know the battery voltages and resistance values, leaving the current values as the unknowns. Solve this system of three equations for the three unknown currents i_1, i_2, and i_3 (your result will still be in terms of the other problem variables). If you have studied methods for solving systems of equations in a math class, feel free to use one of those techniques. If not, you can proceed using the "substitution method," where you solve two of the equations for an unknown and plug these into the third equation.[9]

---

[8] Like with the third voltage loop, the second junction gives you redundant information. For n total loops or junctions in a circuit, you only need $n - 1$ equations, as the nth equation will not provide any new information.

[9] To prevent yourself from going in (mathematical) circles when using substitution, we would suggest the following approach. Look at the two loop equations and notice that they share a common current: in our case, Equations (23.14) and (23.15) both contain the current i_2. Based on this, solve the two loop equations for the *uncommon* currents. In other words, solve Eq. (23.14) for i_1 and Eq. (23.15) for i_3. Finally, substitute these two results into Eq. (23.16), leaving only a single unknown current (i_2). You can solve this for i_2, and then plug this result back into the rearranged loop equations to find solutions for i_1 and i_3.

You have found the "solution" to this circuit. For a particular set of battery and resistor values, you would simply plug in the numbers to determine the current in each branch of the circuit. It is worth reemphasizing that the directions of the currents and voltage loops are arbitrary. You could just as easily have chosen different directions for any of them and you still would have gotten the correct answer.

For example, let's say you had chosen i_2 to point up in the middle branch instead. Your loop and current equations would have been slightly different due to this choice, and your final result for i_2 would have had the same magnitude but opposite sign as compared to before. A negative sign for a current simply tells you that the current flows in the opposite direction of your assumption; in other words, your "wrong" original guess gets automatically corrected during the process!

23.9 (OPTIONAL) VERIFYING KIRCHHOFF'S RULES EXPERIMENTALLY

In this section we will build a circuit to experimentally verify the results from Kirchhoff's rules in Activity 23.8.2. To build this circuit you will need the following items:

- 3 different carbon resistors (e.g., 39 Ω, 75 Ω, 100 Ω)
- 4 D-cell batteries, 1.5 V
- 1 3-D-cell battery holder
- 1 D-cell battery holder
- 1 protoboard
- 1 multimeter
- Assortment of small lengths of #22 wire (for use with the protoboard)

It is common to design and wire "prototype" circuits on a device called a *protoboard*. A protoboard has hundreds of little holes that can accommodate small-diameter wires. In the protoboard model shown in Fig. 23.17, these holes are electrically connected in vertical columns of 5 near the middle. The top of the protoboard has two horizontal rows of 40 connected holes. There is a similar arrangement at the bottom.

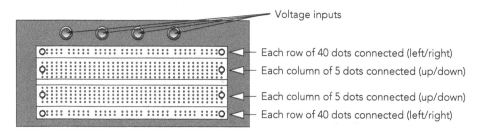

Fig. 23.17. A small protoboard showing the wiring scheme.

Usually, one connects the voltage inputs to the long rows of connected dots toward the outside of the circuit; these rows can then serve as power supplies. We will construct and test out the Kirchhoff's rules circuit of Fig. 23.16 using a protoboard.

23.9.1. Activity: Testing Kirchhoff's Rules

a. Start by measuring the actual values of the three resistors and the two sets of batteries using a multimeter. List the results below.

Measured voltage of battery ΔV_1:

Measured voltage of battery ΔV_2:

Measured resistance of resistor R_1:

Measured resistance of resistor R_2:

Measured resistance of resistor R_3:

b. Return to your theoretical solutions for the unknown currents from Activity 23.8.2. Plug in the measured values of the voltages and resistances for the appropriate variables to find numerical results for the three currents. Write the theoretical results for the currents below.

c. To wire up the circuit shown in Fig. 23.16 on the protoboard, we need to understand the protoboard wiring scheme from Fig. 23.17. Although there are many legitimate ways to connect the leads, a possible configuration is shown in Fig. 23.18. Wire the protoboard to match the connections from the circuit in Fig. 23.16, being sure to put the batteries and resistors in the appropriate locations.

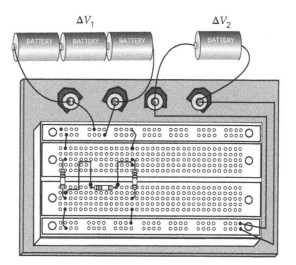

Fig. 23.18. A possible wiring arrangement for the Kirchhoff's rules circuit.

d. We are now ready to measure the current in each branch. After all the effort to wire the protoboard, it would be a shame to have to disconnect branches to insert an ammeter. Fortunately, there is another way! Use a *voltmeter* to measure the potential difference across each resistor, and then use Ohm's law (along with the measured resistance) to determine the current. Do this, listing your results in the table below and comparing your experimental currents with the theoretical ones.

Resistor	Measured R (Ω)	Measured ΔV (volts)	Measured $i = \Delta V/R$ (amps)	Theoretical i (amps)
R_1				
R_2				
R_3				

23.10 PROBLEM SOLVING

Let's return to the circuit of Activity 23.6.2 (shown again in Fig. 23.19).

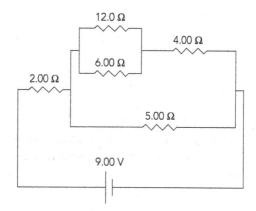

Fig. 23.19. Circuit from Activity 23.6.2.

23.10.1. Activity: Power in a Complex Circuit

Use your results from Activity 23.6.2, along with the concepts of equivalent resistance and Ohm's law, to find the *power dissipated* in the 4 Ω resistor.

Next, consider the following circuit containing a battery and five *identical* light bulbs.

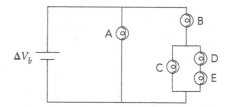

Fig. 23.20. Circuit containing five identical light bulbs.

23.10.2. Activity: Bulb Brightness

a. *Rank* the bulbs according to brightness from brightest to dimmest. Briefly explain your ranking. **Hint**: This can be done using only current arguments, as in Unit 22.

b. Calculate the *voltage drop* across bulb C in terms of the battery voltage V_b (there are multiple ways to do this). Please show your work. **Note**: Although not necessary, if you find it helpful you can assume each bulb has a resistance R.

c. Assume bulb D burns out. Determine what happens to the brightness of each bulb *as compared to its brightness in part* (a). For example, "brighter than before," "dimmer than before," "goes out," etc. Briefly explain your reasoning in each case.

Bulb A (brighter/dimmer/same/out):

Bulb B (brighter/dimmer/same/out):

Bulb C (brighter/dimmer/same/out):

Bulb E (brighter/dimmer/same/out):

Finally, consider the circuit shown in Fig. 23.21.

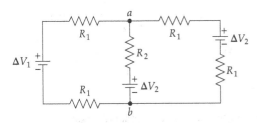

Fig. 23.21. A circuit containing multiple batteries. Assume $\Delta V_1 = 3.0\,\text{V}$, $\Delta V_2 = 6.0\,\text{V}$, $R_1 = 2.0\,\Omega$, and $R_2 = 4.0\,\Omega$.

23.10.3. Activity: Multiple Batteries

a. Determine a set of equations that would allow you to solve for the current through resistor R_2. Keep everything in terms of the problem variables for now (i.e., *don't* plug in numbers yet).

b. Plug in the voltage and resistance values given in the caption to find a numerical result for the current through resistor R_2.

c. What is the potential difference between points a and b?

UNIT 24: CAPACITORS AND *RC* CIRCUITS

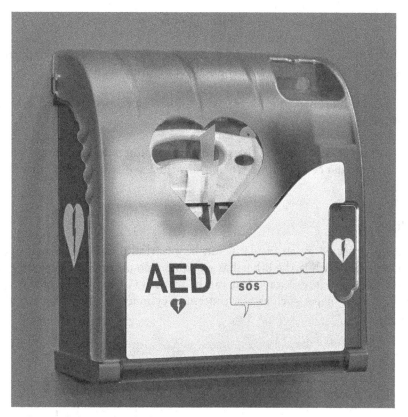

Baloncici/Shutterstock

Automated external defibrillators (AEDs) such as the one shown in this photo are installed in many public gathering spaces. AEDs are used for emergency situations related to a person's heartbeat and have the ability to analyze and treat conditions such as ventricular fibrillation by sending electrical currents through the body. One of the key components of an AED is a capacitor, which consists of two conductors separated from each other by a dielectric material. Capacitors come in all shapes and sizes and are found in most modern electrical devices. The name capacitor suggests a capacity to hold something, but what? How does a capacitor behave in a circuit, including circuits in which the current changes with time? When you complete this unit, you should be able to answer these questions.

UNIT 24: CAPACITORS AND *RC* CIRCUITS

OBJECTIVES

1. To define capacitance and discover how the capacitance of two conducting plates is related to their area and separation distance.

2. To determine how capacitance changes when capacitors are wired in series or parallel.

3. To discover how both the charge on a capacitor and the electric current in a circuit change with time when a capacitor is placed in a circuit with a resistor.

24.1 OVERVIEW

Any two conductors separated by an insulator can be electrically charged so that one conductor has a net excess positive charge, while the other has an equal amount of excess negative charge; such an arrangement is called a *capacitor*. A capacitor can be made up of two blobs of metal, or it can have any number of symmetric shapes such as two concentric cylinders, or two parallel, rectangular plates (see Fig. 24.1).

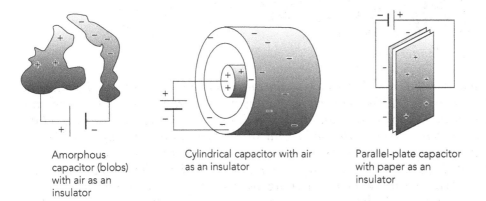

Amorphous capacitor (blobs) with air as an insulator

Cylindrical capacitor with air as an insulator

Parallel-plate capacitor with paper as an insulator

Fig. 24.1. Some different capacitor geometries. In each case, a battery is used to put the excess charges on the two conductors.

The particular shape of any given capacitor is not critical for understanding the basic principles, so we will focus our attention on the parallel-plate capacitor. Such a capacitor is relatively straightforward to construct and analyze. The circuit symbol for a capacitor is a pair of parallel lines as shown in Fig. 24.2.[1]

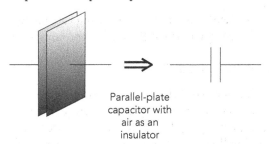

Parallel-plate
capacitor with
air as an
insulator

Fig. 24.2. The circuit diagram symbol for a capacitor is designed to look like a parallel-plate capacitor.

Capacitors are used widely in circuits and are contained in most modern electronic devices. Although many interesting properties of capacitors come in the operation of alternating current circuits, we will limit our present study to the properties of the parallel-plate capacitor and the behavior of capacitors in circuits with constant voltage sources like those you have been constructing in the last couple of units.

FUNDAMENTALS OF CAPACITORS

24.2 THE PARALLEL-PLATE CAPACITOR

The typical method for transferring equal and opposite excess charges to a capacitor is to use a voltage source such as a battery to induce a potential difference between the two conductors. Electrons will then flow off one conductor (leaving excess positive charges) and onto the other until the potential difference between the two conductors is the same as that of the voltage source.[2] In general, the amount of excess charge needed to reach the potential difference of the battery will depend on the size, shape, and location of the conductors relative to each other. The *capacitance* of a given capacitor is defined mathematically as the ratio of the amount of excess charge $|q|$ on either one of the conductors to the magnitude of the potential difference $|\Delta V|$ across the two conductors:

$$C \equiv \frac{|q|}{|\Delta V|} \tag{24.1}$$

Thus, capacitance has units of Coulombs per volt in SI units and, by definition, C will always be positive.[3]

---

[1] For some capacitors, it matters which side is positively charged and which is negatively charged. In this case the symbol typically includes plus and minus signs on the two plates, like what you see for a battery symbol. We will address these situations as they arise.

[2] At first glance it might be surprising that current will flow at all because the circuit is not "closed" in the traditional sense (there is a gap between the two plates of the capacitor). We'll consider the details of this process shortly.

[3] Note that the plates on the capacitor will always have equal amounts of opposite charge. On one plate there will be a positive charge $|q|$ and on the other plate there will be a negative charge $-|q|$.

To complete the activities in this section, you will need the following items:

*   A few capacitors (assorted collection)
*   2 pieces aluminum foil, 12 cm × 12 cm
*   1 textbook
*   1 multimeter (with capacitance measuring capability)
*   2 insulated wires (stripped at the ends, approximately 6″ long)
*   1 ruler
*   1 Vernier caliper (optional)

Figure 24.3 shows a basic circuit for connecting a capacitor to a battery. When the switch is open, there is no (excess) charge on either plate (assuming the capacitor began uncharged). Upon closing the switch, charges begin to flow onto the plates and will continue until the voltage across the two conductors is the same as that of the battery. From Eq. (24.1) we can see that for a fixed voltage battery, the net excess charge found on either plate is proportional to the capacitance of the pair of conductors. We will use our knowledge of electrostatics and circuits to understand why this happens.

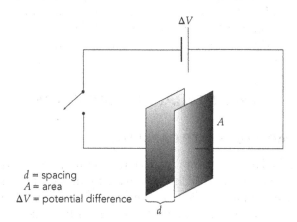

Fig. 24.3. A parallel-plate capacitor will have a potential difference ΔV across the plates once the switch is closed. The plates have area A and are separated by a distance d.

24.2.1. Activity: Predicting Dependence on Area and Separation

a. Consider two identical metal plates of area A separated by a gap of non-conducting material (such as air) with thickness d. The plates are connected in a circuit with a battery and switch as shown in Fig. 24.3. After the switch is closed, what happens to the net charge on each plate? Do the excess charges on one plate of a charged capacitor interact (exert forces) with the excess charges on the other plate? If so, how? Could this help explain why charges are able to flow onto the plates, even though the circuit is not "closed" in the usual sense?

b. Is there a limit to the amount of excess charge that can be put on a plate? Explain, referring to Eq. (24.1).

c. Imagine increasing the areas of both plates while the potential difference remains fixed. Use *qualitative* reasoning to predict how the amount of excess charge on each plate should change as the areas of the plates increase. Briefly explain.

d. Now imagine moving the plates farther apart (increasing d) while the potential difference remains fixed. Predict how the amount of excess charge on each plate should change as the spacing between the plates increases. Briefly explain.

Measuring Capacitance

The unit of capacitance, named after Michael Faraday, is the farad (F) and is equal to one coulomb/volt. As we have mentioned, a coulomb is a large amount of charge, which makes one farad a large amount of capacitance. Typically, capacitances are expressed in units of mircrofarads, nanofarads, or picofarads. In addition, the labels on capacitors may use somewhat confusing notation by having the symbol U (or even m) represent micro (μ) when labeling a capacitor (see table):

Measurements of Capacitance
microfarad: 10^{-6} F = 1 μF (sometimes labeled 1 UF, 1 uF, or even 1 mF)
nanofarad: 10^{-9} F = 1 nF
picofarad: 10^{-12} F = 1 pF

There are several types of capacitors used in electric circuits, including disk capacitors, foil capacitors, and electrolytic capacitors. Before starting the next activity, you should examine some typical capacitors to see the different styles.

In the next activity you will make a parallel-plate capacitor out of two rectangular sheets of aluminum foil separated by pieces of paper. A textbook works well as the separator for the foil since you can slip the two foil sheets between any number of sheets of paper and weigh the book down with something heavy and non-conducting, like another textbook. **Note**: Insert short wires into the capacitance slots of your multimeter as probes. When you measure the capacitance of your "parallel plates," be sure the aluminum foil pieces are arranged carefully so they don't touch each other and "short out."

24.2.2. Activity: How Capacitance Depends on Area and Separation

a. Devise a way to measure how the capacitance depends on *either* the foil area or the separation between foil sheets. If you hold the area constant and vary the separation, record the dimensions of the foil so you know the area. Alternatively, if you hold the distance constant, record its value. Take at least four data points in either case. **Note**: Your instructor may split up the class so that half the groups vary one parameter while the other half varies the other parameter. Describe your methods and record your data (including units). Create a plot showing how the capacitance varies with your parameter (either area or separation). Sketch or print out the graph of your results.

b. Is your graph a straight line? If so, fit the graph with the appropriate function and write your fit equation below. If your graph is not a straight line, try to guess the functional relationship it represents and try fitting your data with this function.[4] Describe the results of your fit below, including the value of any constants.

c. What is the function that best describes the relationship between capacitance and area? What about the relationship between capacitance and separation? How do the results compare with your predictions from Activity 24.2.1?

---

[4] If you want to try fitting $1/d$ but your software does not allow this type of fit, you instead can plot C vs. $1/d$ and try fitting that with a straight line.

d. Just to verify that current is not actually flowing from one plate to the other through the pages of the book, use the multimeter to measure the *resistance* of the thickness of pages you used. Can any current flow *through* the pages of your book? **Note**: You should remove the capacitor plates and leads when doing this.

24.3 A MATHEMATICAL EXPRESSION FOR CAPACITANCE

In Unit 20 we used Gauss's law to find the electric field some distance away from a large plane ("sheet") of charge. We found that

$$\vec{E} = \frac{\sigma}{2\varepsilon_0} \text{ away from sheet} \quad \text{(uniform plane of charge)} \tag{24.2}$$

where σ is the surface charge density of the sheet (which can be positive or negative). For positive charge, the field points *away* from the sheet, while for negative charge, the field points *toward* the sheet. (Remember that our result of a constant electric field strength relied on an assumption that the sheet was large compared to the distance away.) We can use this result and the relationship between potential difference and electric field to derive an expression for the capacitance of a parallel-plate capacitor in terms of the area A and separation d of the conducting plates.

Figure 24.4 shows the geometry of the situation, where two plates of a capacitor are separated by a distance d. The bottom plate is assumed to have positive charge on its inner surface, while the upper plate has negative charge on its inner surface. An imaginary Gaussian surface shaped like a rectangular box is drawn such that its top surface is just above the bottom plate and its bottom surface is inside the bottom plate.

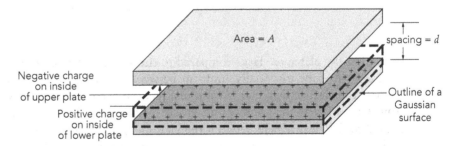

Fig. 24.4. Two plates of a capacitor. The dotted lines outline a rectangular box that serves as the Gaussian surface enclosing positive charge on the bottom plate. Although not shown, the two plates are assumed to extend outward in all directions so that the electric field is essentially uniform between the two plates.

Note: In Unit 20 we assumed the sheet was surrounded by vacuum (or at least air). However, for most capacitors there is a "dielectric material" between the plates, which results in a modification to Gauss's law. In particular, the constant ε_0 is replaced by the constant $\varepsilon = \kappa\varepsilon_0$, where κ ("kappa") represents the dielectric constant of the material separating the two plates ($\kappa \approx 1$ for air).

24.3.1. Activity: Derivation of Capacitance for a Parallel-Plate Capacitor

a. Examine the imaginary Gaussian "box" shown in Fig. 24.4. From Eq. (24.2) we know the magnitude and direction of the electric field through the top surface of the Gaussian box due to (only) the positively-charged plate. In the space below, write down the electric field (magnitude and direction) in the region *between* the plates due to the positively-charged plate. Be sure to incorporate the new constant κ in your expression.

b. Similarly, we could imagine using a Gaussian box to enclose the top, negatively-charged plate. Without going through all the steps of Gauss's law, write down the electric field (magnitude and direction) in the region *between* the plates due to the negatively-charged plate. Once again, be sure to incorporate the new constant κ in your expression.

c. The net, or total, electric field in the region between the plates is given by the superposition of these two fields. Use your results from parts (a) and (b) to determine the net field between the two plates.

d. Assuming the plates are large compared to the spacing, we expect the electric field to be essentially uniform between the two plates. And because the electric field points from one plate to the other, we know from Unit 21 that the potential difference between the two plates is related to the electric field between the plates as $|\Delta V| = |\vec{E}|d$, where d is the distance between the two plates. Use this fact, your result from part (c), and the definition of capacitance to find an expression for C in terms of the plate area A, plate separation d, dielectric constant κ, and ε_0. **Hint**: Remember that σ is a charge per unit area.

You should have found that for our parallel-plate capacitor,

$$C = \frac{\kappa \varepsilon_0 A}{d} \quad \text{(parallel-plate capacitor)} \quad (24.3)$$

e. Use one set of your values for area and spacing from the measurements you made in Activity 24.2.2 to calculate a value of *C* for your textbook parallel-plate capacitor. Assume that the dielectric constant κ for paper is 3.5. How does the calculated value of *C* compare with the value you measured?

f. You probably found that the value of capacitance for your homemade capacitor was quite small (in farads). Let's say you want to use the same arrangement to make a capacitor that has a capacitance of $C = 1$ F. How long, in meters, would each side of your (square) sheet of foil need to be to reach this capacitance, assuming $\kappa = 3.5$ and $d = 1$ mm? Show your calculations.

g. Your capacitor would make a mighty large circuit element! How could it be made physically smaller and still have the same value of capacitance? You may want to examine the collection of sample capacitors for some ideas.

24.4 CAPACITOR NETWORKS

Like with resistors, we are interested in determining the equivalent capacitance for series and parallel combinations of capacitors. For the next set of activities, you'll need the following equipment:

- 4 D-cell batteries, 1.5 V
- 1 D-cell battery holder
- 1 3-D-cell battery holder
- 2 capacitors, 0.1 µF (or similar)
- 2 large-capacitance capacitors, 0.47 F (or similar)
- 6 alligator clip wires
- 1 SPST switch
- 1 multimeter (with capacitance measuring capability)

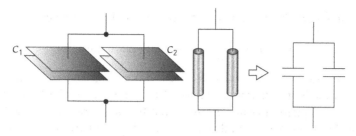

Fig. 24.5. Two capacitors wired in parallel. Left: physical picture of two parallel-plate capacitors. Middle: two cylindrical capacitors. Right: circuit symbols.

24.4.1. Activity: Capacitance for a Parallel Arrangement

a. We start with two identical capacitors wired in parallel as shown in Fig. 24.5. Use physical reasoning to predict the equivalent capacitance for a pair of capacitors wired in parallel. Explain your reasoning. **Hint**: What is the effective area of two parallel-plate capacitors wired in parallel?

b. Next, wire up two identical capacitors (0.1 μF or similar) in parallel and measure the total (equivalent) capacitance of the pair using a multimeter. How does your measured value compare to your prediction? **Note**: It can be tricky to correctly wire and measure two capacitors in parallel. The "tops" of the two capacitors should be connected together, as should the "bottoms" of the two capacitors. If your capacitors are *polarized* (have positive and negative sides labeled), be sure to wire them accordingly.

c. Based on your results, predict a general equation for the equivalent capacitance of a parallel combination as a function of the two capacitances C_1 and C_2.

Next, consider two capacitors wired in series, as shown in Fig. 24.6.

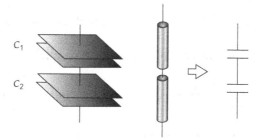

Fig. 24.6. Two capacitors wired in series. Left: physical picture of two parallel-plate capacitors. Middle: two cylindrical capacitors. Right: circuit symbols.

24.4.2. Activity: Capacitance for a Series Arrangement

a. Use physical reasoning to predict the equivalent capacitance of a pair of capacitors wired in series. Explain your reasoning. **Note:** This is more subtle than the parallel case. It helps to realize that the length of the wire connecting the two inner plates (bottom of C_1 to top of C_2) is not important. In fact, this wire could be shortened to essentially zero length without changing the capacitance. Thinking about the two capacitors connected by a wire of zero length might help you predict the equivalent capacitance.

b. Next, wire up two identical capacitors (0.1 μF or similar) in series and measure the total (equivalent) capacitance of the pair using a multimeter. How does your measured value compare to your prediction?

c. Based on what you reasoned and measured, can you predict a general equation for the equivalent capacitance of a series combination as a function of the two capacitances C_1 and C_2? **Note:** Again, this is not as straightforward as for the parallel combination. You may find that thinking about combinations of resistors provides some guidance.

d. How do the mathematical relationships for series and parallel capacitors compare to those of resistors?

You should have found that capacitors in parallel add like resistors in series:

$$C_{eq} = C_1 + C_2 + C_3 + \cdots + C_n \quad \text{(capacitors in parallel)} \qquad (24.4)$$

Likewise, capacitors in series add like resistors in parallel:

$$\frac{1}{C_{eq}} = \frac{1}{C_1} + \frac{1}{C_2} + \frac{1}{C_3} + \cdots + \frac{1}{C_n} \quad \text{(capacitors in series)} \qquad (24.5)$$

It's worth making another comparison to resistors to help us understand how combinations of capacitors act in circuits. For resistors in parallel, the potential differences are the same, while the currents can be different (the current is "divided up" among the resistors). For resistors in series, the current through each is the same, while the potential differences can be different (the total voltage gets "divided up" among the resistors).

For capacitors, it is the voltage and *charge* on the capacitor that we consider. Just like with resistors (or any other element), capacitors in parallel will each have the same potential difference. However, the charges can be different and are determined by the capacitances via the relationship $q = C\Delta V$ (the total charge is "divided up" among the capacitors). For capacitors in series, the charge on each is the same, while the potential differences can be different (the total voltage is "divided up" among the capacitors).[5]

Networks of Capacitors

When dealing with networks of capacitors, we can use the rules for combining series and parallel capacitors to find the equivalent capacitance of an entire network. This equivalent capacitance can then be used to find the total charge stored, and subsequently the individual charges and potential differences for each capacitor. As in the case of resistors, one wants to begin as deep into the circuit as possible where it's clear which capacitors are in series or parallel.

24.4.3. Activity: Equivalent Capacitance and Charges

a. Consider the circuit shown in Fig. 24.7. Find the equivalent capacitance of the capacitor network. Then, redraw the final circuit with the

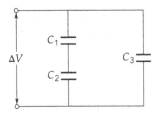

Fig. 24.7. A circuit with three capacitors. Assume $\Delta V = 3.0$ V, $C_1 = 2.0$ μF, $C_2 = 5.0$ μF, and $C_3 = 7.0$ μF.

---

[5] These relationships are relatively easy to prove using what we already know about circuits and the expression $q = C\Delta V$.

battery and a single, equivalent capacitor representing the equivalent capacitance.

b. Although the equivalent capacitor you drew in the simplified circuit is not real (there is no actual capacitor with this exact value in the circuit), we can think of it as if it were a real capacitor to help us understand how the entire circuit behaves. For example, use the equivalent capacitor of the circuit to determine the *total charge* stored in the network of capacitors.

c. Use the known value of ΔV to find the charge stored on capacitor C_3.

d. Use your answers to parts (b) and (c) to determine the charge on capacitor C_1. What then is the charge on capacitor C_2?

A Capacitance Puzzle

Suppose two identical capacitors are hooked up to 1.5 V and 4.5 V batteries in two separate circuits (as shown in Fig. 24.8). They are each unhooked from their batteries (without being discharged) and connected to each other in parallel. We are interested in determining the final potential difference across the pair.

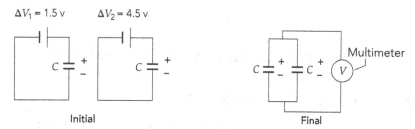

Fig. 24.8. Two identical capacitors are charged by different batteries before being connected together in parallel.

24.4.4. Activity: Solving the Puzzle

a. What do you predict the final potential difference will be across the two capacitors in parallel? Briefly explain.

b. Try to use our capacitor equations to calculate what will happen. **Hints**: What do you know about the initial charge on each capacitor? What do you know about the final sum of the charges on the two capacitors (assuming no discharge takes place)? Use this information to find the final voltage across the pair.

c. Now try the experiment. Set up the initial circuits and charge the two capacitors. Carefully remove the batteries and wire up the two capacitors in parallel. Then measure the final potential difference across the parallel combination. **Warnings**: Be sure to connect the two positive sides together and the two negative sides together! Also, avoid touching the leads of the capacitors with your fingers or stray wires while modifying the circuit; it's very easy to accidentally discharge a capacitor without realizing it!

d. Was your prediction correct? If not, how do you need to modify your reasoning to correctly predict the final voltage?

CHARGE BUILDUP AND DECAY IN CAPACITORS

24.5 INTRODUCTION TO *RC* CIRCUITS

In the next section we'll examine the time dependence of the voltage across a charged capacitor when it is placed in a circuit with a resistor (an "*RC* circuit"—*R* for resistor and *C* for capacitor). Before performing these measurements, we will make some qualitative observations of capacitor behavior so that we have a better understanding of what happens when a capacitor charges and discharges. For the observations in this section, you will need the following:

- 3 D-cell batteries, 1.5 V
- 1 3-D-cell battery holder
- 1 #14 bulb
- 1 #48 bulb
- 2 large-capacitance capacitors, 0.47 F (or similar)
- 1 multimeter
- 6 alligator clip wires
- 1 SPST switch

Qualitative Observations

By using a light bulb as a resistor and one or more large-capacitance capacitors, we can "see" what happens to the current flowing through the bulb as a capacitor is either charged (by a battery) or discharged.

24.5.1. Activity: Qualitative Observation of an *RC* Circuit

Fig. 24.9. A #14 bulb (rounded).

a. Wire a single-loop circuit that connects a rounded #14 bulb (see Fig. 24.9) with a 0.47 F capacitor, a switch, and a 4.5 V battery (three D-cells in series). Draw a circuit diagram of your setup. **Important**: If your capacitor is polarized with terminals marked positive and/or negative, be sure the negative side of the capacitor is closest to the negative side of the battery! Close the switch and describe or sketch what you observe (assuming the capacitor is initially uncharged).

b. Can you think of a way to make the bulb light up again *without* the battery in the circuit? Try connecting the capacitor, bulb, and switch in a way that makes the bulb light up (at least for a short time). Describe your observations and draw a circuit diagram showing the setup.

c. Describe or sketch the approximate brightness of the bulb as a function of time when it is placed in a circuit with a charged capacitor *without* the battery, as in part (b). Assume the switch is closed at time $t = 0$ in your graph.

d. Repeat the charging and discharging processes again to get a semi-quantitative picture of what happens. While doing so, use the multimeter to measure the voltage across the capacitor during the process of both charging and discharging. Use your voltage measurements to briefly explain why the light bulb changes brightness as time progresses.

e. What happens when *more capacitance* is added to the circuit? Wire up a second capacitor in an appropriate way to *increase* the total capacitance. Then repeat the charging/discharging process. What do you observe?

f. What happens when *more resistance* is put in the circuit? Remove the second capacitor so that you are back to the original configuration. Then replace the round #14 bulb with the oblong #48 bulb (see Fig. 24.10), which has a larger resistance. Repeat the charging/discharging process and describe what do you observe.

Fig. 24.10. A #48 bulb (elongated).

24.6 QUANTITATIVE MEASUREMENTS OF AN *RC* SYSTEM

Our next task is to perform a more quantitative study of an *RC* system. We want to measure the potential difference across a capacitor as a function of time $\Delta V_C(t)$ as it is being charged or discharged through a resistor. Our goal is to determine the mathematical relationship that best describes the potential difference as a function of time as the capacitor charges (or discharges).

For the activities in this section, you will need:

- 3 D-cell batteries, 1.5 V
- 1 3-D-cell battery holder
- 1 capacitor, approx. 5000 μF (or 0.47 F)
- 1 resistor, approx. 1.0 kΩ (or 10 Ω)
- 1 SPDT switch
- 1 data-acquisition system
- 1 electronic voltage probe
- 6 alligator clip wires
- 1 capacitance meter for C $\geq$5000 μF (optional)

We know a bulb's resistance is temperature dependent and rises when it is warm up due to a current. For these quantitative studies we'll use a 1.0 kΩ resistor in place of the bulb while charging a 5000 μF capacitor (or a 10 Ω resistor with

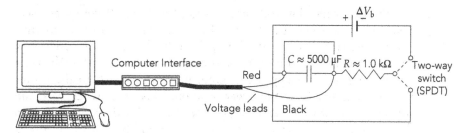

Fig. 24.11. An *RC* circuit that switches between charging and discharging a capacitor.

a 0.47 F capacitor). Wire up the circuit shown in Fig. 24.11 using a single-pole, double-throw (SPDT) switch. This switch will allow you to quickly change from a situation in which the battery is charging the capacitor through a resistor to one in which the capacitor is allowed to discharge through a resistor. The voltage leads from your computer interface should be used to measure the potential difference across the capacitor. If your capacitor is polarized, be sure it is oriented appropriately.

Hint: The circuit in Fig. 24.11 can be a bit tricky to wire correctly. You may find it helpful to pick one place to start and then follow the wires around. For example, the negative side of the battery should be connected by a wire to one side of the SPDT switch. The other side of the SPDT switch is connected to one side of the capacitor (as well as the positive side of the battery). Be sure to ask if you have questions!

24.6.1. Activity: *RC* Charging and Discharging

a. Assume the capacitor is initially uncharged. By examining Fig. 24.11, briefly explain how you think the circuit will behave when the switch is moved to the *upper* position. Sketch a *predicted* graph of the potential difference across the capacitor as a function of time, assuming the switch is connected to the upper position at time $t = 0$.

b. Now assume the switch is moved to the *center* position (*not* connected to either side). Predict what should happen (if anything) to the voltage across the capacitor with the switch in this position.

c. Now consider moving the switch to the *lower* position. How do you think the circuit will behave when the switch is moved to the lower position? Sketch a *predicted* graph of the potential difference across the capacitor as a function of time, assuming the switch is connected to the lower position at time $t = 0$.

d. Set the data collection software to record at least 20 points per second for a total time of at least 120 seconds. Make sure the capacitor is fully *dis*charged before beginning. Start the data collection and move the switch to the upper position so the capacitor charges. After the voltage has essentially stopped changing, move the switch to the center position, and finally to the lower position. Move the switch back and forth a couple of times, allowing plenty of time for the voltage to stop changing before you change the switch position. Sketch the recorded $V_C(t)$ graph and identify the regions on the graphs where the capacitor is either charging or discharging. In these regions, does the graph agree with your predictions from part (a) (charging) and part (c) (discharging)?

e. We now wish to fit the discharging data. (You can either select an appropriate region of the data you just took, or you can collect new data only for discharging.) Using the software, try to find a functional form that fits the data well. **Note**: Be sure to only fit the data *after* the switch was closed; if the data collection started before the switch was fully closed, you should carefully select the appropriate region of data. Sketch or print out your plot with the fit. Write the equation of your fit, including the values of any constants. **Hint**: We have seen this same function appear in other experiments!

The Theoretical *RC* Decay Curve

In the previous activity, you should have found that an exponential decay curve of the form

$$\Delta V_C(t) = Ae^{-\alpha t} + B$$

fits the data quite well (with $B \approx 0$). Both A and α are constants provided with the fit (your fitting software might use different notation for these two constants). This curve has the same mathematical form as the cooling curve we encountered in the study of heat and temperature when a warm object was left to cool in constant-temperature surroundings. For the cooling-curve experiments, the exponential decay was the result of the rate of cooling being proportional to the temperature difference between the object and its surroundings. In the following activity we will perform a theoretical analysis of the capacitor discharge process to see why it also follows an exponential decay.

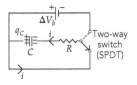

Fig. 24.12. An initially-charged capacitor discharging through a resistor R.

24.6.2. Activity: Derivation of the Discharge Curve

a. We consider the same circuit as in Activity 24.6.1 with the capacitor starting out fully charged before the switch is moved to the lower (discharge) position. Figure 24.12 depicts the circuit, where an excess charge of q_C is on the capacitor plates. Assuming the switch has been moved to the lower position, use Kirchhoff's Voltage Loop Rule around the bottom loop of the circuit to relate the voltage ΔV_C across the capacitor to the current i through the resistor and the resistance R. **Reminder**: The Voltage Loop Rule of Unit 23 can also be used on circuits containing capacitors!

b. Next, use the definition of capacitance in Eq. (24.1) to replace ΔV_C in your result to part (a) with q_C and C. Rearrange your equation to solve for the current i.

At this point, you should have an equation that reads $i = q_C/RC$, where q_C is the amount of charge on the capacitor. But we also know that current is simply the rate of charge flow:

$$i \equiv \frac{dq}{dt}$$

where this q is the charge in the wire. We can substitute this expression to replace the current in your result to part (b) and end up with an equation where the only time-dependent variable is the charge q.

There is one tricky aspect we need to consider due to the fact that the two q's are not exactly the same: a positive current through the resistor in Fig. 24.12 ($i > 0$) results from charge *leaving* the capacitor ($dq_C/dt < 0$). To account for this sign difference, we write the current as $i = -dq_C/dt$, where q_C represents the charge on the capacitor. Carrying out this substitution leads to the following expression:

$$\frac{dq_C}{dt} = -\frac{1}{RC}q_C \qquad (24.6)$$

This is a *differential equation* for the charge $q_C(t)$ on the capacitor, which tells us that the rate of change of charge on the capacitor is equal to a (negative) constant times the amount of charge on the capacitor. This is very similar to the expression for Newton's law of cooling in Eq. (16.3)! As before, we wish to show that our fit equation in Activity 24.6.1 "satisfies" Eq. (24.6). Because our fit equation involves the voltage across the capacitor as a function of time, the first step will be to rewrite our fit equation in terms of the charge on the capacitor as a function of time.

c. Rewrite your fit equation from Activity 24.6.1(e) in the space below, leaving the two constants as variables (do not plug in actual numbers). Then use the definition of capacitance to replace $\Delta V_C(t)$ with $q_C(t)$ and C. Move everything except $q_C(t)$ to the right side of the equation. This is our fit equation in terms of the charge on the capacitor, or $q_C(t)$.

d. Take the time derivative of the equation for $q_C(t)$ you found in part (c). Your result is the left side of Eq. (24.6).

e. Now substitute your expression for $q_C(t)$ from part (c), as well as your expression for dq_C/dt from part (d), into Eq. (24.6). Then solve for the unknown constant α in terms of the other problem variables.

You should have found that your fit equation *does* satisfy the differential equation, but only if the constant α is equal to $1/RC$. In other words, the left side of Eq. (24.6) equals the right side if we assume that $q_C = Ae^{-t/RC}$. In fact, if we plug $t = 0$ into the equation we see that the constant A must be equal to the initial charge on the capacitor. Moreover, because our fit equation is known to agree with experiment, we now have confidence that Eq. (24.6) provides a valid description of the physical situation.

In summary, we have found both experimentally and theoretically that the voltage across, and charge on, a discharging capacitor are given by

$$\Delta V_C(t) = \Delta V_0 e^{-t/RC} \quad \left(\text{discharging capacitor}\right) \tag{24.7}$$

$$q_C(t) = q_0 e^{-t/RC} \quad \left(\text{discharging capacitor}\right) \tag{24.8}$$

where ΔV_0 is the initial voltage on the capacitor, and q_0 is the initial charge on the capacitor before the discharge begins.

Note that either one of these equations can be used to describe the discharge—they are simply related through the definition of capacitance! Additionally, because current is the rate of change of charge, part (d) of the previous activity shows that the current will also decrease exponentially in time for a discharging capacitor: $i(t) = i_0 e^{-t/RC}$. Finally, notice that the time for the voltage (or charge or current) to decrease depends on both R and C, with larger values leading to longer times. We will quantify this idea shortly.

24.6.3. Activity: Putting in the Numbers

a. Consider the expression for the voltage across a discharging capacitor given by Eq. (24.7), which incorporates the fact that the experimental fit parameter α found in Activity 24.6.1 is equal to $1/RC$. We can use this information to experimentally determine the value of the capacitance and compare it to the rated value written on the outside of the capacitor. Use a multimeter to measure the resistance R of the resistor, being sure to remove the resistor from the rest of the circuit first. Then use this measured resistance and the value of α from your fit equation to determine an experimental value of C. How does this compare to the rated value listed on the capacitor?[6]

b. Finally, verify that the constant A from your fit corresponds to the initial voltage across the capacitor. **Note**: The initial voltage is the voltage across the capacitor at the time corresponding to the *beginning of your fit*.

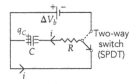

Fig. 24.13. An initially-charged capacitor discharging through a resistor R.

A Summary of *RC* Decay

The last few activities have introduced a lot of new information about capacitors in *RC* circuits. Therefore, before moving on we finish this section with an activity summarizing what we've seen for a discharging capacitor. Figure 24.13 shows the circuit again for convenience.

24.6.4. Activity: Explaining Discharging Qualitatively

a. Assume the battery voltage in our circuit is ΔV_b and that the capacitor starts fully charged. What is the potential difference ΔV_C across the capacitor *just before* the switch is moved to the discharge position?

---

[6] If you have a meter capable of measuring capacitances in excess of 5000 μF, you can also perform a direct measurement of the capacitor. Standard multimeters tend not to be able to measure such large capacitances.

b. Is the current through the resistor a maximum or a minimum *just after* the switch is flipped at time $t = 0$? Does current ever flow *across the gap between the plates* of the capacitor? Briefly explain.

c. How is the magnitude of the voltage across the resistor related to the magnitude of the voltage across the capacitor during the discharge process?

d. Sketch four plots below: $\Delta V_C(t)$, $q_C(t)$, $i(t)$, and $\Delta V_R(t)$.

CHARGING AND TIME SCALES IN *RC* CIRCUITS

24.7   A CHARGING CAPACITOR

In the previous activities we assumed the capacitor started charged and focused on the discharge process. Here we consider the reverse situation: an uncharged capacitor C being charged by the battery with current flowing through a resistor R. Figure 24.14 depicts the situation, where the switch is now moved to the upper position for charging. The qualitative and quantitative considerations of this situation are very similar to that of a capacitor discharging. We won't go through all the details, but instead use what we learned to make some predictions. You may find it useful to think about Kirchhoff's Voltage Loop Rule in the following activity.

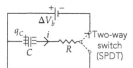

Fig. 24.14. An initially-discharged capacitor is charged by a battery through a resistor R.

24.7.1. Activity: Charging Capacitor Behavior

a. Assuming the capacitor starts fully discharged, what is the voltage across the capacitor *immediately after* the switch is moved to the upper position?

b. Do you expect the current to be large or small when the switch is first closed? Why? What should happen to the current as time progresses? **Hint**: How is the current related to the voltage across the resistor?

c. If you wait long enough, what value should the voltage across the capacitor reach?

d. What do you expect a plot of $\Delta V_C(t)$ to look like? Similarly, what do you expect a plot of $q_C(t)$ to look like? Sketch your predicted graphs below, assuming the switch is moved to the upper position at time $t = 0$.

Because $\Delta V_C = 0$ initially, the magnitude of the initial voltage across the resistor immediately after the switch is moved must be the same as that of the battery. This implies the current starts out at a maximum (rapid charging at the beginning). As charge begins to build up on the capacitor, the voltage across the capacitor increases, leaving a smaller voltage across the resistor. This means the current decreases, and the capacitor charges more slowly. Eventually the potential difference across the capacitor is equal to that of the battery, and the charging stops.

Mathematically, the voltage and charge on the capacitor can be described by the equations[7]

$$\Delta V_C(t) = \Delta V_B \left(1 - e^{-t/RC}\right) \quad \text{(charging capacitor)} \tag{24.9}$$

$$q_C(t) = q_0 \left(1 - e^{-t/RC}\right) \quad \text{(charging capacitor)} \tag{24.10}$$

where ΔV_B is the potential difference across the battery and q_0 is the maximum charge on one of the capacitor plates when fully charged: $q_0 = C\Delta V_B$.

24.8 *RC* DECAY TIMES

Recall that the equation describing how a capacitor discharges through a resistor is given by

$$\Delta V_C(t) = \Delta V_0 e^{-t/RC}$$

---

[7] Just like the discharging capacitor equation was analogous to an object subject to Newton's Law of Cooling, a charging capacitor is analogous to a cold object warming up in a constant-temperature environment!

Similarly, as we just learned, the equation describing a charging capacitor also involves an exponential of the form $e^{-t/RC}$. The product RC has units of time and is called the *RC time constant* (or just the *time constant*) of the circuit. As we will see, the time constant determines how long it takes for the capacitor to charge or discharge.

24.8.1. Activity: Theoretical *RC* Decay Times

a. We just said the product RC has units of time. Consider a discharging capacitor at a time $t = RC$. Show mathematically that at a time equal to RC, the potential difference ΔV_C across the capacitor drops to 36.8% of its initial value.

b. Another convenient time is that describing the *half-life* of the discharge, or the time is takes for the voltage across the capacitor to drop to *half* of its initial value: $\Delta V_C = \Delta V_0/2$. One typically denotes the time corresponding to the half-life as $t_{1/2}$. Use this definition to determine the half-life $t_{1/2}$ in terms of R and C.

Hopefully, you found that the half-life is given by

$$t_{1/2} = RC \ln 2 = 0.693\, RC \quad \text{(half-life of } RC \text{ circuit)} \qquad (24.11)$$

In words: at a time of $t_{1/2} = 0.693\, RC$ the voltage across the capacitor should be exactly half of where it started.[8]

Note: Although we only discussed the RC time constant and half-life with respect to a discharging capacitor, the same principle holds for the charging expressions as well.

---

[8] You may have considered similar ideas in the exponential decay of radioactive substances (e.g., carbon dating). The half-life in a radioactive process is the time it takes for half of the atoms to decay, or the time when the radioactive atoms are half-depleted (see Unit 28).

Measuring the Half-Life

In the next activity we will record the change in voltage as a capacitor discharges. You will need the following items to complete this activity:

- 1 data-acquisition system
- 1 electronic voltage probe
- 1 multimeter (with capacitance and resistance measuring capability)
- 3 D-cell batteries, 1.5 V
- 1 3-D-cell battery holder
- 1 capacitor with largest capacitance that can be measured by your capacitance meter
- 2 different resistors with resistances such that the product *RC* is in the range 5–20 seconds
- 1 SPDT switch
- 6 alligator clip wires

You should set up the charging/discharging circuit as shown in Fig. 24.11. For this activity you will change the value of the resistance and check the validity of Eq. (24.11) by experimentally measuring $t_{1/2}$.

24.8.2. Activity: Measuring the Half-life

a. Use a capacitance meter to measure the capacitance of your capacitor. Measure it multiple times and if the values differ use the average of the readings. Write the average value below. You will want to use this value going forward (as opposed to the rated value noted on the capacitor).

$$C =$$

Fig. 24.15. The time for decay of charge from a capacitor in an *RC* circuit is sometimes called "relaxation time."

b. Use an ohmmeter to measure the resistance of both resistors. Write the values below. You will want to use these values going forward (as opposed to the rated values given by the color code).

$$R_1 =$$
$$R_2 =$$

c. Describe the method you are using to verify Eq. (24.11). Include a sketch of your experimental circuit, a description of what data you will collect, and how you will use this data to find an experimental value of $t_{1/2}$.

d. For *each* of the resistors, perform the experiment and determine the value of the half-life for the discharge. For *one* of the experiments, sketch or print out a graph showing $\Delta V_C(t)$ with notes showing how you determined the value of $t_{1/2}$ from the graph.

> Experimental $t_{1/2}$ with $R_1 =$
>
> Experimental $t_{1/2}$ with $R_2 =$

e. Finally, use your measured capacitance and resistance values from parts (a) and (b) to calculate the *expected* half-life from Eq. (24.11) (for both resistance values). How close are these values to the half-life you measured experimentally in part (d)?

24.9 PROBLEM SOLVING

A Capacitor and Resistor Network

Consider the circuit in Fig. 24.16. As indicated, assume the capacitor arrangement is initially charged to a voltage of 9 V before the switch is closed.

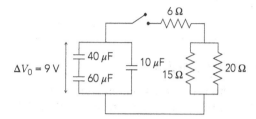

Fig. 24.16. Arrangement of capacitors and resistors. The capacitor network is originally charged to a potential of 9 V.

24.9.1. Activity: Capacitor and Resistor Network

a. Find the equivalent capacitance of the arrangement of capacitors. Show your steps clearly, including any equations that you use.

b. Find the equivalent resistance of the arrangement of resistors. Show your steps clearly, including any equations that you use.

c. Redraw the circuit using your equivalent values from parts (a) and (b). Then determine the *time* at which the current through the 6 Ω resistor has decreased to *half* of its initial value from when the switch is first closed.

Construction Zone Warning Lights

The flashing lights in highway construction zones often operate using what is known as an *oscillator circuit*. The circuit is shown in Fig. 24.17, where the light L is a neon (or other) gas tube that is wired in parallel with a capacitor. The circuit works as follows:

- The gas in the light is a good insulator normally, so this branch of the circuit has essentially infinite resistance when the light is off. This means that no current can flow through the light, and the capacitor charges up through the resistor R.
- Once the voltage across the two leads in the gas tube reaches the *breakdown voltage* V_{on}, the electric field between the contacts is strong enough to ionize the gas. This creates a spark like we saw with the van de Graff generator, and the gas gives off the familiar orange glow due to the spark.
- The ionized gas acts like a very good conductor (essentially a wire), and so the capacitor rapidly discharges through the gas tube. The capacitor voltage quickly drops until it reaches a value of V_{off}, at which point the electric field is too weak to keep the spark active. At this point the light turns off, and the capacitor starts to charge once again. This entire process then repeats over and over again.

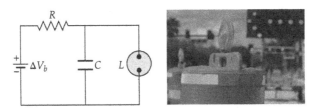

Fig. 24.17. Left: Circuit for controlling a flashing construction light. Right: Picture of a typical construction light (Condor 36/Adobe Stock Photos).

The voltage across the capacitor (and the gas tube) as a function of time looks like that shown in Fig. 24.18. When the circuit is first turned on, the capacitor begins to charge. It passes through V_{off} on the way up, and once it reaches V_{on} for the first time, the lamp flashes. The voltage then rapidly falls to V_{off}, at which point it starts to climb back up again toward V_{on} (when the lamp flashes again).

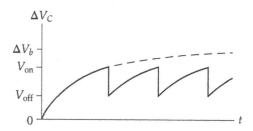

Fig. 24.18. Voltage across the capacitor as a function of time for the flashing light. Dashed line shows what the voltage would look like if the gas tube were not present.

24.9.2. Activity: Flashing Lights

a. Suppose we construct the circuit with the following values: $\Delta V_b = 90$ V, $V_{off} = 30$ V, $V_{on} = 70$ V, $R = 75$ kΩ, and $C = 10$ μF. (V_{off} and V_{on} are determined by how the lamp is constructed.) Calculate the time it takes for the voltage across the capacitor to reach V_{off} the *first* time it charges.

b. Similarly, calculate the time it takes for the voltage across the capacitor to reach V_{on} the *first* time it charges.

c. Based on your answers and the graph in Fig. 24.18, how much time is there between each flash? **Remember**: The discharge time through the spark is essentially zero, so it can be ignored.

d. Determine the number of times the lamp flashes each second (the flash *rate*).

UNIT 25: ELECTRONICS

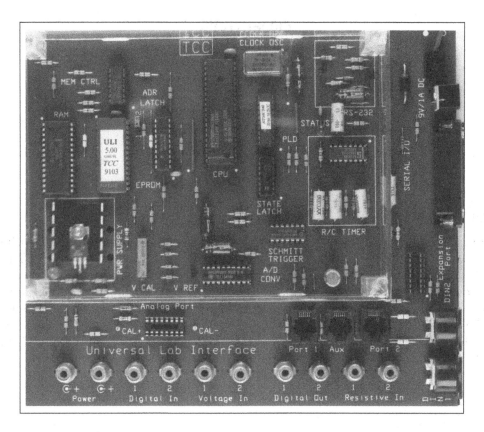

This photograph shows one of the very first Universal Laboratory Interfaces designed for use in the Workshop Physics program. It consists of a printed circuit board and electrical components that allow one to acquire electrical signals from laboratory equipment and send them to a computer. In addition to the familiar looking resistors and capacitors on the board, there are long black plastic elements known as integrated circuits. These integrated circuits each contain thousands of tiny circuit elements including semiconducting devices known as transistors (along with resistors and capacitors). In this unit you will learn some basic electronics, as well as how to use an oscilloscope to monitor time-dependent voltages in circuits.

UNIT 25: ELECTRONICS

OBJECTIVES

1. To understand how an oscilloscope can be used to display changes in voltage over time and use it to investigate alternating-current waveforms.

2. To gain experience with some fundamental logic operations used in digital electronics.

3. To construct a circuit that operates like a stopwatch!

25.1 OVERVIEW

Electronics is a vast field that deals with the information contained in electrical signals, such as those that produce sound from a loudspeaker, a TV picture, or memory states in a computer. This unit is intended to provide you with a brief introduction to electronics, including some of the devices used and the ways you can transform electrical signals by designing and constructing circuits. We start with the oscilloscope as a fundamental instrument used to measure time-dependent voltages. Then we explore some of the digital logic components that provide the basis for modern electronics. Finally, we will use a variety of electronic components to build a stopwatch. Although we can only cover the basics, hopefully the activities in this unit will give you a taste of the field of analog and digital electronics.

THE OSCILLOSCOPE AND TIME-DEPENDENT SIGNALS

25.2   INTRODUCTION TO THE OSCILLOSCOPE

Determining how potential differences (voltages) change over time is a very common diagnostic in many scientific laboratories, and the *oscilloscope* (or *scope* for short) is the standard instrument for such measurements. Oscilloscopes are typically stand-alone devices (see Fig. 25.1), but they can also be integrated into a desktop computer. In the most common configuration, an oscilloscope receives one or more electrical signals as inputs and displays graphs of the voltages as functions of time.

Fig. 25.1. A stand-alone digital oscilloscope.

Modern oscilloscopes use digital circuitry to measure the voltage and then display the graph. Although you will likely make measurements using such a digital oscilloscope, we begin this section by considering the original, analog version of the oscilloscope because the design and operation of this device connects nicely with material from earlier units. Your instructor may have an analog scope to show as a demonstration.

The heart of an analog oscilloscope is an evacuated glass tube known as a cathode ray tube, or CRT for short.[1] As shown in Fig. 25.2, it contains a source of electrons (an "electron gun") and multiple sets of metal deflection plates that can hold electrical charge. The input voltage signal we wish to measure is connected to one set of deflection plates, which acts like a capacitor. As we've learned, when there is a potential difference across the two plates of a capacitor, the plates become charged and there is an electric field between them. The electron gun sends a beam of negatively-charged electrons between the two plates, and the beam is deflected due to the electric field. By measuring how far the electron beam is deflected, we can determine the potential difference across the plates (and hence the voltage of the input signal).

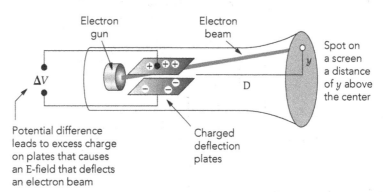

Fig. 25.2. A simplified view of the CRT in an oscilloscope. Only one set of parallel plates is shown; there is another pair oriented in the perpendicular direction.

---

[1] The CRT was also the primary component of analog televisions.

In the following activity, we analyze the motion of an electron moving perpendicular to a uniform electric field between two charged plates. Your instructor may have the following equipment available:

- 1 demonstration oscilloscope (optional)

Motion of an Electron in a Uniform Electric Field

In this section we investigate how the motion of an electron in an electric field leads to a relationship between the input potential difference and the electron's deflection that allows an oscilloscope to display changes in voltage graphically. We will also see how the electron's trajectory is similar to motion we have seen before.

We'll assume the parallel-plate capacitor has a uniform electric field in the region between the plates. As we saw in Unit 21, the magnitude of this electric field can be written as

$$E = \frac{|\Delta V|}{d} \tag{25.1}$$

where ΔV is the potential difference between the two plates, and d is the spacing between the plates. The direction of the field is determined by a vector pointing away from the positively-charged plate and toward the negatively-charged plate.

Let's consider the motion of an electron as it traverses the electric field between the two plates of the oscilloscope. A side view of the situation is shown in Fig. 25.3. Based on the diagram we know the directions of the field and velocity, so we'll only concern ourselves with magnitudes in what follows.

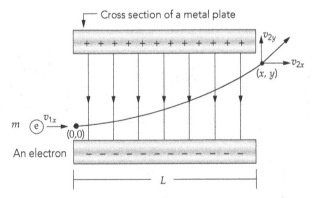

Fig. 25.3. An electron with velocity component v_{1x} enters the region between the two plates of an oscilloscope that has a potential difference across them. The electron is deflected due to the electric field and emerges with final velocity components v_{2x} and v_{2y}.

25.2.1. Activity: Analyzing Motion in a Uniform Electric Field

a. An electron with charge magnitude q enters the region with an electric field $\vec{E}$. Write down an expression for the magnitude of the force F on the electron in terms of the magnitude of the electric field.

b. As shown in Fig. 25.3, the electric field points in the negative y-direction. In what direction does the electron accelerate? Assuming the electron has mass m, use Eq. (25.1) and Newton's second law to find the magnitude of the electron's acceleration in the region between the plates in terms of the charge q, mass m, potential difference ΔV, and separation d. If the electric field between the plates is uniform in space and constant in time, what can we deduce about the acceleration?

c. If the electron enters the region with an x-component of velocity v_{1x}, what is the x-component of the velocity when it exits the region?

d. You should have concluded that the electron accelerates in the (positive) y-direction while the velocity in the x-direction remains constant. Does this remind you of another type of motion we have studied? What is the shape of the path the electron will take as it passes through the plates?

e. Assume the plates have a length L as shown in Fig. 25.3. Use your result from part (b) to determine the y-component of the velocity v_{2y} when the electron exits the region between the plates. Express your result in terms of the charge q, mass m, potential difference ΔV, separation d, plate length L, and initial x-component of velocity v_{1x}.

f. All the parameters in your result to part (e) are constant except ΔV, the voltage applied across the two plates, which means that v_{2y} at the edge of the plates is proportional ΔV. After passing through the plates, the electron leaves the influence of the electric field and travels an *additional* distance D before hitting a phosphorescent screen that glows at the spot of impact (see Fig. 25.2). Explain why the displacement of the spot on the screen is a measure of the potential difference across the plates.

Graphing the Input Voltage

Activity 25.2.1 shows how the displacement of the spot on the screen is a measure of the voltage applied to the plates. Thus, at any point in time the vertical position of the spot is a measure of the input voltage to the oscilloscope. Note that when the input voltage goes negative, the charges on the plates switch signs, deflecting the spot in the opposite direction (down instead of up). This allows one to measure both positive and negative input voltages, and with a proper calibration, the deflected position provides a quantitative measure of the input voltage in volts.

    We are halfway to explaining how the oscilloscope manages to plot the input voltage $V(t)$; what remains is to understand how the scope can plot different input voltages at different times.

25.2.2. Activity: Graphing V(t)

Although not shown in Fig. 25.2, the oscilloscope has a second set of plates oriented perpendicular to the first set. This set of plates can produce an electric field in the *horizontal* direction, the strength and direction of which are controlled by internal circuits in the oscilloscope. Can you think of how the oscilloscope might control the horizontal deflection so that the phosphorescent screen shows a graph of the input voltage as a function of time?

To create a graph of how the input voltage changes over time, the second pair of plates are provided with a "sweep" (or "sawtooth") potential difference that varies in time as shown in Fig. 25.4. The resulting electric field in the horizontal direction causes the spot where the electrons strike the phosphor screen to move smoothly from left to right at a fixed rate for the duration of the sawtooth wave. When combined with the input voltage, the spot on the screen maps input voltages at different times to different horizontal positions on the screen, creating a

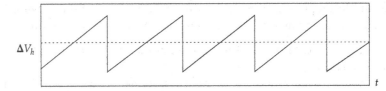

Fig. 25.4. Sawtooth voltage applied to the horizontal deflection plates in the oscilloscope.

graph of $V_{in}(t)$. The beam then quickly returns to the left edge for another sweep across the screen. The speed with which the electrons move along the time axis is called the time base and can be adjusted using knobs on the oscilloscope.

25.3 GENERATING AND MEASURING TIME-DEPENDENT SIGNALS

To observe how an oscilloscope works in action, we need a voltage source that changes in time. As we saw in earlier units, a battery acts as a constant-voltage source, so instead we use a device known as a *function generator* (sometimes called a *signal generator* or *wave generator*). A function generator can create various time-dependent electrical waveforms, or "voltage waves." In this section we investigate such time-dependent signals using a function generator and an oscilloscope.

You will need the following equipment:

- 1 function generator
- 1 oscilloscope (or other source to measure time-dependent signals)
- 1 speaker
- 1 BNC cable (or other method to connect the function generator to the speaker and scope)

Using a Function Generator

A basic function generator is pictured in Fig. 25.5 (yours may look somewhat different). The various knobs and buttons control such parameters as the type of

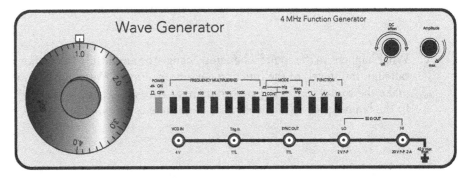

Fig. 25.5. An example function generator.

waveform, the amplitude of the waveform, the frequency of the waveform, etc. We'll start by setting up the function generator to produce a sine wave output:

1. Plug in and turn on the function generator. Turn *off* the DC OFFSET (if applicable).
2. Set the function generator to produce a sinusoidal wave by selecting the sine wave output (typically a button with a small sine wave above it).
3. Set the function generator so that the frequency of the sine wave is around 250 Hz. This is typically accomplished in one of two ways. If your function generator has a frequency dial and multiplier button, you'll want to set the frequency dial to 2.50 and the multiplier to 100 Hz (creating an output of $2.50 \times 100\,\text{Hz} = 250\,\text{Hz}$). If your function generator instead has frequency range buttons and coarse/fine knobs, you should set the range button to a value close to 250 Hz (e.g., 500 Hz), and then use the coarse/fine knobs until the output reads 250 Hz.
4. Turn *down* the output amplitude all the way.
5. Proceed with Activity 25.3.1.

25.3.1. Activity: Sound from a Function Generator

a. Connect the output of the function generator to a speaker using either a BNC cable or clip leads. Slowly turn up the output amplitude until you hear a tone coming from the speaker. Describe what you hear when you have the sine wave output set at 250 Hz.

b. What happens to the sound when you change the frequency of the function generator output? In the language of music, what does frequency correspond to?

c. What happens to the sound when you change the amplitude (sometimes called output level) of the function generator output? **Warning**: Do *not* make the sound too loud; you can damage the speaker (or your ears)! In the language of music, what does amplitude correspond to?

d. Turn down the amplitude until the tone is just audible. Describe what happens to the sound when you select the triangle or square wave outputs instead of the sine wave. Return the function generator to the sine wave setting when you are done.

The previous activity uses a speaker to convert the electric voltage wave into a sound wave. As we'll see in Unit 27, the changing voltage causes the speaker to vibrate back forth, producing a sound wave in the air (and which your ears sense). The frequency of the wave corresponds to the *pitch* of the sound, while the amplitude controls the *volume*. By choosing a different output function (triangle or square wave instead of sine), the *timbre* (pronounced "tamber"), or *character*, of the sound is changed. In the next activity, we use an oscilloscope to visualize the waveforms.

Using an Oscilloscope

We are now ready to use the oscilloscope to measure and plot the time-dependent signal coming from the function generator. Figure 25.6 shows the front of a generic oscilloscope. The details of the set-up, menus, dials, etc. will depend on the exact type of scope you are using, so we'll attempt to keep the instructions as general as possible. Your instructor may have more specific guidelines to follow. In either case, you will want to start by plugging in and turning on the oscilloscope at your lab station. When you first turn it on, the scope will likely go through some sort of warm-up or initialization routine, which may take a minute or so.

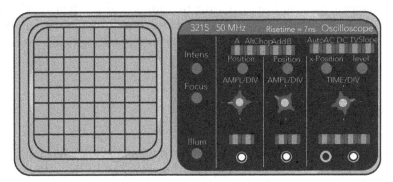

Fig. 25.6. The front of a generic oscilloscope. Although your scope may look different, the functionality is generally the same.

25.3.2. Activity: Basic Oscilloscope Controls

a. If your oscilloscope has a "Default Setup" button or selection, push that to set the scope to its default configuration. (If not, proceed directly with the next step.)

b. Use a BNC cable to connect the output of the function generator to the channel one (CH 1) input of the oscilloscope. You should hopefully see a sine wave across the oscilloscope screen. If not, you may need to turn up the output amplitude of the function generator slightly. You may also need to adjust the "triggering" of the oscilloscope; this is what tells the scope when to start plotting the input signal. The "Auto Trigger" function of the scope works best for our purposes. Before moving on to part (c), be sure you see a sine wave signal on the scope and ask for assistance if needed.

c. The oscilloscope shows you the voltage as a function of time over a short time interval (the duration of the interval depends on the scope settings). The horizontal scale represents time while the vertical scale represents volts. Adjust the knob that controls the *time scale* for the scope (typically labeled something like "Time/Div" or "Horizontal Scale"). What changes about the graph when you adjust the time scale on the scope? Is the actual signal changing, or just how we are viewing it (how "zoomed in" we are)? Leave the setting so that you see approximately five complete oscillations of the sine wave across the screen.

d. Adjust the knob that controls the *horizontal position* on the scope (typically labeled "x-position" or "Horizontal Position"). What changes about the graph when you adjust the horizontal position on the scope?

e. Next, adjust the knob that controls the *voltage scale* (typically something like "Volts/Div," "Ampl/Div," or "Vertical Scale"). What changes about the graph when you adjust the voltage scale on the scope? Is the actual signal changing, or just how we are viewing it? Leave the setting so that the wave fills most of the screen.

f. Finally, adjust the knob that control the *vertical position* on the scope (typically labeled "*y*-position" or "Vertical Position"). What changes about the graph when you adjust the vertical position on the scope? Leave the setting so the sine wave is centered on the middle of the screen.

The oscilloscope plots the input voltage as a function of time, or $V(t)$. As shown in the previous activity, you can easily control the duration of time that is plotted (zooming in or out on the horizontal axis), as well as the voltage scale of the graph (zooming in or out on the vertical axis). You can also adjust the position of the graph on the screen, either left-and-right (in time) or up-and-down (in voltage). Note that none of these changes affect the actual input signal; the input signal is controlled by the function generator, which we didn't touch! The oscilloscope knobs simply control how we *view* the signal. We are now ready to investigate the usefulness of the oscilloscope for measuring and analyzing input signals.

25.3.3. Activity: Measuring Time-Dependent Waveforms

a. Let's use the oscilloscope to determine the *period* of the sine wave coming from the function generator. There are multiple ways to do this, and we'll start with the most basic. The number of seconds per box (or "time per division") on the scope screen is shown on the screen (for a digital scope) or on the time knob (for an analog scope). Find this number and then estimate the number of divisions one complete wave covers. Use this to determine the period of the wave, or the time it takes for the wave to go through one complete oscillation.

b. How does the period you found in part (a) compare with the period calculated using the frequency reading from the function generator? **Hint**: You'll need to remember (or look up) how period and frequency are related for a wave.

c. With a digital oscilloscope, one can do better than counting boxes. If you have a digital scope, you can use the "Cursors" menu to position vertical lines one period apart on the waveform. The scope will then tell you how far apart in time these vertical lines are from each other (thereby giving you the period). In fact, for simple time-dependent signals like a sine wave, the oscilloscope can do this all automatically. The "Measure" (or similar) menu will let you select a channel to analyze, and the scope will show you its measured period, frequency, amplitude, etc. If you are using a digital oscilloscope, try out both the Cursors and Measure functions. How do the results compare to your estimate using the number of boxes?

d. Next, adjust the frequency knob on the *function generator* while watching the oscilloscope. What happens to the waveform on the scope? Is the wave itself changing or just how we are viewing it?

e. Now adjust the amplitude knob on the *function generator* while watching the oscilloscope. What happens to the waveform on the scope? Again, is the wave itself changing or just how we are viewing it?

f. Try experimenting with the "DC Offset" control on the function generator and the "AC/DC" (or AC Coupling) button on the oscilloscope. Explain how these controls in combination appear to affect the oscilloscope display. Return both of these to their original setting when done.

g. Finally, switch over to the triangle wave and square wave outputs of the function generator. Sketch all three waveforms in the boxes below.

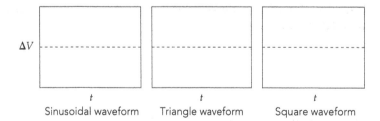

ΔV

Sinusoidal waveform   Triangle waveform   Square waveform

25.4   THE MICROPHONE AND SOUND WAVES (OPTIONAL)

In Activity 25.3.1 we saw that sending a time-varying voltage to a speaker generates a sound wave. Then in Activity 25.3.3 we visualized this time-varying voltage using an oscilloscope. It turns out that one can make the reverse process happen as well: sending a sound wave into a speaker generates a time-varying voltage, which can then be visualized using an oscilloscope. (You probably already knew this, although you likely used the more common term "microphone" for this type of speaker!)

The electrical output from an unamplified microphone is quite small, so we'll use the computer for data acquisition instead of the oscilloscope. To do the experiments in this section, you will need:

Andrii Zastrozhnov/
Shutterstock

- 1 data-acquisition system
- 1 microphone (for data-acquisition system)
- 1 source of music (e.g., personal cell phone or stand-alone MP3 player)

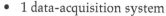

25.4.1. Activity: The Microphone

a. Open up your data-acquisition software on the computer and connect the microphone. Set the software to collect data for approximately 0.05 s. Collect data while you sing a note (or hum) into the microphone. Given the short duration of collection, you will likely want to start signing *before* you start collecting and continue singing until *after* you see the waveform on the screen. Sketch the waveform you see below. How does the waveform compare to the sine wave of the previous section?

b. Trying singing at a different pitch—what changes about the waveform? Trying having someone else sing a note instead—do the traces look different?

c. Now play a sample of music and use the microphone to record/display the electrical waveform the represents this music. After viewing it once or twice, change the software to collect data for a longer duration of time (e.g., 0.7 s). How does the waveform representing the music compare to that for one person singing a note (and to the sine wave)? Why do you think this might be?

The microphone is essentially a small speaker that converts an incoming sound wave into an electrical signal (which is then viewed on the computer or oscilloscope). This is the reverse process of a regular speaker, which takes an electrical signal and converts it into a sound wave. Both processes typically rely on the same underlying physics known as electromagnetic induction, a concept we will study in Unit 27.

You probably noticed that the waveform for a person singing a note has more structure than a single-frequency sine wave. It turns out that most instruments, including the human voice, produce multiple frequencies (known as *harmonics*), even when playing (or singing) a "single note." The waveform representing a musical song likely contains not only harmonics, but also multiple instruments playing simultaneously. The resulting complex waveform as a function of time is due to the combination of many different frequencies in the sound wave.

DIGITAL LOGIC AND THE STOPWATCH

25.5 DIGITAL SIGNALS AND LOGIC GATES

The waveforms we generated in the previous sections are *analog* signals. The term analog in electronics refers to a circuit where voltages can vary continuously over time and take on any value. For example, the sine wave sweeps through all possible voltages between its minimum and maximum values during each period. Most modern electrical devices, such as computers and cell phones, are built with circuits that use digital rather than analog electronics. Let's explore the meaning of the term "digital" when it comes to electronics.

Digital electronics involves circuits where voltages can take on only a finite number of discrete values. In fact, for most circuits there are only two possible values: they are either "on" or "off," where 0 V is the off value and 5 V is a typical on value. The square wave we saw in the previous sections could be used to represent a type of digital signal. For example, by shifting the DC Offset, the square wave can be made to jump between 0 V and 5 V only. In such digital circuits, the voltage will never take on the value of, say, 1.37 V; any potential difference greater than a cutoff value will be interpreted as the "on" state, whereas any potential difference less than the cutoff value will be the "off" state. This feature makes digital circuitry remarkably insensitive to stray variations or "noise" in the signal and is one reason computers and other modern digital devices rely on digital electronics.[2]

Most digital circuits are based on transistors, and we will not attempt to describe the physics behind their operation in this class. Instead, we will use digital circuits packaged into devices known as integrated circuits with the goal of understanding how digital circuit elements behave and what can be done with them (rather than analyzing why they behave as they do).

---

[2] While only two voltage values are typically used in digital electronics, these two can be used *represent* a wide variety of other values using binary numbers. As one common example, music is often stored and transmitted as a digital file that is later converted to analog before reaching a speaker.

Integrated Circuits and Protoboards

Integrated circuits, or "ICs," are devices in which many tiny transistors, along with diodes, capacitors, and resistors, are connected together with thin metal films to make elaborate circuits. ICs are often bundled together and put in a small, single package that has a row of connectors on each side. Two of these connectors are typically used to provide electrical power to the IC, while the rest serve as inputs and outputs.

The first types of ICs we will use are known as *logic gates*. These ICs perform logical operations and when combined can execute more advanced functions. As we'll see, the inputs and resulting outputs for logic gates can be summarized in something called a *truth table*.

When manually constructing circuits containing ICs, one uses a device known as a *protoboard*, which allows one to easily wire together circuit elements in a temporary manner using numerous small, press-fit wire connections ("holes"). A diagram of a typical protoboard is shown in Fig. 25.7. The lines along the rows and columns show which press-fit holes are connected together underneath the board. A wire put into any hole along one of these rows or columns will be electrically connected to every other hole in that row/column.

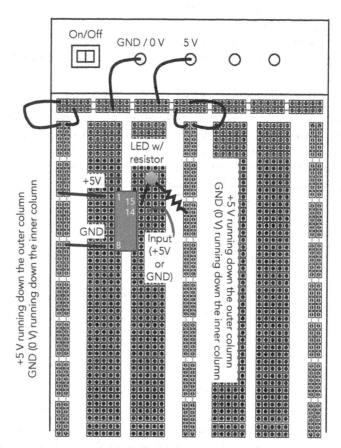

Fig. 25.7. A diagram of a protoboard showing the underlying electrical connections, along with the wiring for a typical IC.

The protoboard of Fig. 25.7 has an internal power supply and provides connections for both "ground" (0 V) and 5 V. It is typical to connect each of these outputs to one of the top two rows of wire connectors (sometimes called "rails"). In the wiring scheme of Fig. 25.7, every press-fit connection in the very top row is at 5 V, while every connection in the second row is at ground. These two rails are then connected as shown to some of the vertical columns running down the protoboard. These columns provide easy access to either 5 V or ground at just about any point on the protoboard.

An IC (represented by the rectangle with numbers on it) is placed into the protoboard straddling one of the gaps running down the board. The connectors in each of the short, horizontal rows on each side of the IC are wired together under the protoboard, allowing one to have multiple wires in contact with each input or output of the IC. As shown in Fig. 25.7, pins 1 and 8 on this particular IC are used to provide power: 5 V is connected to pin 1 and ground is connected to pin 8. In the next activity, we will wire up an IC.

The NOT and AND Gates

We'll first explore two basic digital ICs in preparation to undertake a more extensive digital electronics project––the building of a digital stopwatch. For the next few activities, you will need the following items:

- 1 NOT (Inverter) gate IC (4049)
- 1 AND gate IC (4081)
- 1 LED with current-limiting resistor
- 1 powered protoboard (with +5 V) (or an unpowered protoboard with batteries)
- Assorted small lengths of wire for the protoboard

The pin designations for a 4049 Inverter are shown on the left side Fig. 25.8 (the right side of Fig. 25.8 depicts the schematic circuit element for an Inverter). Pins 1 and 8 provide power, while pins 2 through 7, 9 through 12, and 14 through 15 contain six independent "NOT" (or Inverter) gates (pin 13 is unused). We will use pins 14 and 15 to start, where pin 14 is the input to the NOT gate and pin 15 is the output.

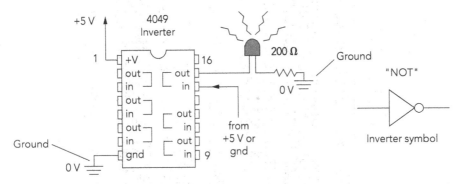

Fig. 25.8. Left: Pin diagram for a 4049 Inverter IC. The LED on the output is used to read off the output state. Right: Circuit diagram symbol for a NOT gate.

We will provide either 5 V or ground to the input, while the output is connected to a basic *light emitting diode* (LED), which then passes through a resistor to ground. The LED provides a simple method to indicate whether the output potential is "on" (+5 V) or "off." (0 V). If the LED is on, the output is +5 V; if it is off, the output is 0 V. **Important**: Be sure to have a current-limiting resistor between the LED and ground to prevent burning out the LED.

Note that there are two ground connections (0 V) in Fig. 25.8 that use the ground symbol ⏚. The electrical ground is a common reference point for many elements in a circuit; all points in a circuit diagram that are marked with a ground symbol are actually connected together (e.g., under the protoboard). This common point is designed to have a large capacity for holding charge so that its potential does not change.[3]

25.5.1. Activity: Truth Table for the NOT Gate

a. Place the 4049 Inverter IC onto your protoboard as shown in Fig. 25.7 (and using the pin diagram of Fig. 25.8). Supply power to the IC with pins 1 and 8. Connect the LED to pin 15 (the output of one of the six Inverter circuits). **Note**: The LED is likely directional; the flat side of the base of the LED bulb should be connected to the point of lower potential (in this case, to ground). Use a wire to connect either ground or +5 V to input pin 15 using the powered rails running down the protoboard. When you have ground going to the input, is the LED on or off? What if you have +5 V going to the input?

b. Use your results from part (a) to complete the "truth table" shown below. A *truth table* lists the output of a logical circuit in terms of the inputs. For a single-input gate like the Inverter, there are only two possible digital input states: "0" which represents ground (also referred to as OFF or LO), and "1" which represents +5 V (also referred to as ON or HI).

NOT Truth Table	
Input	Output
0	
1	

---

[3] The name "ground" comes from the fact that most electrical grounds eventually connect into the actual ground of Earth (which has an essentially limitless capacity to hold charge).

c. Why do you think this circuit is called a NOT (or Inverter) gate? What does it accomplish?

___

We saw that the Inverter gate flips, or inverts, the state of the signal; a 0 becomes a 1 while a 1 becomes a zero. In logic language, this is called a NOT gate (or a NOT operation). We next consider a logical AND gate. The 4081 AND contains four identical AND gates, where each gate has two inputs and one output (see left side of Fig. 25.9). Although not shown, we will again use an LED to monitor the output state.

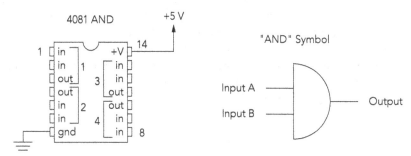

Fig. 25.9. Pin diagram for a 4081 AND IC. Note that the ground is specified only by the ground symbol. Right: Circuit diagram symbol for an AND gate.

___

25.5.2. Activity: Truth Table for the AND Gate

a. Place the 4081 AND IC onto your protoboard and power it using the pin diagram of Fig. 25.9. Choose one of the four AND gates contained in the IC (e.g., pins 8–10) and use two wires to connect either ground or +5 V to the input pins. Connect an LED to the appropriate output pin (with the other side of the LED going through a resistor to ground). For a logic gate with two independent inputs (each of which can be 0 or 1), how many possible input combinations are there?

b. Check the output for each combination of inputs and complete the truth table below for the AND gate.

AND Truth Table		
Input A	Input B	Output
0	0	
0	1	
1	0	
1	1	

c. Why do you think this circuit is called an AND gate?

d. Imagine that input A is held at 0 V, while input B is a continuous string of pulses (a "pulse train"). That is, the voltage of B alternates between 0 V and 5 V in time. *Predict* what the output will look like as a function of time. (You might find it helpful to draw the string of pulses for input B and the output below that.)

e. Now imagine input A is held at 5 V while input B is again a string of pulses. *Predict* what the output will look like as a function of time. In comparing your answers for parts (d) and (e), in which configuration do the pulses at input B "pass through" the gate?

25.6 THE DIGITAL STOPWATCH

For the remainder of this unit, we will construct a digital stopwatch using several ICs and some resistors and capacitors. In addition to the 4081 AND gate, we will need a 555 Timer, two 4026 Counters, and two 7-segment displays. We proceed in steps, first exploring how these other ICs operate before combining them. You will need the following items to construct the stopwatch:

- 1 555 Timer (for example, Sylvania ECG955)
- 2 base-10 counters (4026)

- 2 seven-segment display elements (3057)
- 1 AND gate (4081)
- 1 potentiometer, 10 kΩ (a variable resistor)
- 6 resistors, 1 kΩ
- 1 electrolytic capacitor, 10 μF
- 1 capacitor, 0.1 μF
- 1 protoboard with +5 V power
- Assorted small lengths of wire for use with protoboard
- 1 microswitch
- 1 oscilloscope
- 1 multimeter with ohmmeter functionality

The 555 Oscillator

To build our stopwatch, we need to construct a pulse train that can be passed through an AND gate when its input A is HI, but not passed through when input A is LO (just like in parts (d) and (e) of Activity 25.5.2). The pulse train is created using an *oscillator circuit* that spontaneously produces a series of pulses. The oscillator circuit employs a standard "555 Timer" that can generate a pulse train at a frequency determined by the values of the capacitor and resistors connected to it. Figure 25.10 shows how to wire up a 555 Timer circuit.

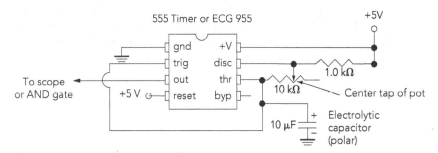

Fig. 25.10. Pin diagram for a 555 Timer IC wired as an oscillator. The 10 kΩ resistor is a potentiometer, or "pot," in which the resistance can be varied.

25.6.1. Activity: Frequency Output of the 555 Timer

a. Wire the circuit as shown in Fig. 25.10 on your protoboard. Observe the output signal of the circuit with the oscilloscope. Sketch and describe what you see.

b. What happens when you change the value of the potentiometer? With the potentiometer at its maximum setting, calculate the frequency of the pulse train. **Hint**: Determine the period of the pulse train on the scope and use the relationship between period and frequency.

c. Adjust the potentiometer so that the frequency of the pulse train is approximately 1 Hz (a period of 1 s).

Counting and Displaying Pulses

The 555 Timer sends out a pulse train with an adjustable frequency/period. If we set the period to be one second, counting the number of pulses tells us how many seconds have elapsed. We therefore need a method for counting pulses. Of course, to be of any practical use, the stopwatch must also display the number of pulses (seconds) that get counted.

For counting we will use the 4026 Counter IC (see Fig. 25.11). The 4026 takes an input clock signal ("clk in" on pin 1) and counts upward in base-10 each time a new pulse arrives in the clock signal. The pulse train output of our 555 Timer will serve as the input clock for the 4026. A single 4026 can only count from 0 to 9, but after ten pulses it produces another output pulse ("÷10 out" on pin 5) that can serve as input clock for a second 4026, thus allowing for counts beyond nine.

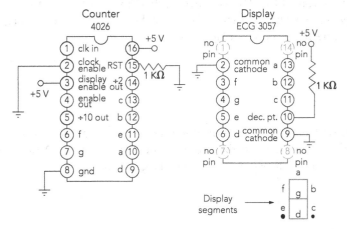

Fig. 25.11. Pin diagrams for a 4026 Counter IC and an ECG3057 Display IC. The display segments corresponding to the various input letters are shown at bottom (some circuits can also display decimal points).

For displaying the count, the 4026 provides seven outputs (letters "a" through "g") that drive an ECG3057 Seven-Segment Display IC. The small diagram at the bottom right in Fig. 25.11 shows the corresponding display segments. For example, if the 4026 is at a count of 5, it outputs HI signals at the proper outputs for the display to read the digital number 5 (segments a, f, g, c, and d). When the 4026 and ECG3057 are connected as in Fig. 25.12, the display produces a numeral that increases by one for each electrical pulse that enters the clock input of the 4026.

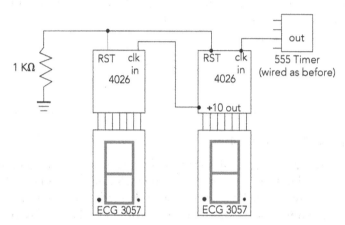

Fig. 25.12. Wiring diagram for two-digit counting with seven-segment displays. The wires between the 4026s and 3057s link a-to-a, b-to-b, etc.

25.6.2. Activity: Counting the Pulses

a. Wire up the two 4026s and two ECG3057s shown in Fig. 25.12. Connect the output of the 555 Timer directly to the clock input of the first 4026 Counter. Connect the ÷10 output of the first 4026 to the clock input of the second 4026. Then connect each 4026 Counter to an ECG3057 Display. **Note**: Don't forget to also make the additional power and grounding connections shown in Fig. 25.11!

b. Describe what you see once you have your counter working properly. This circuit is a two-digit counter; how large a number can it count to? What happens when it gets to its largest number?

c. Try adjusting the potentiometer in your 555 Timer circuit. What happens to the counting rate?

Starting and Stopping the Count

To have a working stopwatch, we obviously need a method for starting and stopping the counting. We'll accomplish this using a "microswitch" (see Fig. 25.13), which has three terminals: common ("C"), normally-open (NO), and normally-closed (NC). The next activity gives you a chance to see how a microswitch functions.

25.6.3. Activity: The Microswitch

a. Use the ohmmeter feature of a multimeter to determine how the common terminal connects to the other two terminals (ideally you can use the "connection test" functionality of the ohmmeter). When the actuator is free (not depressed), which terminal is the common connected to? How can you tell?

Fig. 25.13. A microswitch with the actuator free (top) and depressed (bottom).

b. When the actuator is depressed, which terminal is the common connected to? How can you tell?

A Functioning Stopwatch

We are now ready to combine the oscillator, counter, display, and other elements used in this unit to construct a stopwatch that measures the elapsed time a spring-loaded switch is held in the "on" position. The stopwatch will also need to reset to zero each time so that it is ready to measure a new time interval. A block wiring diagram of the full circuit is shown in Fig. 25.14 (you should have already wired up the 555 Timer, 4026 Counters, and 3057 Displays separately).

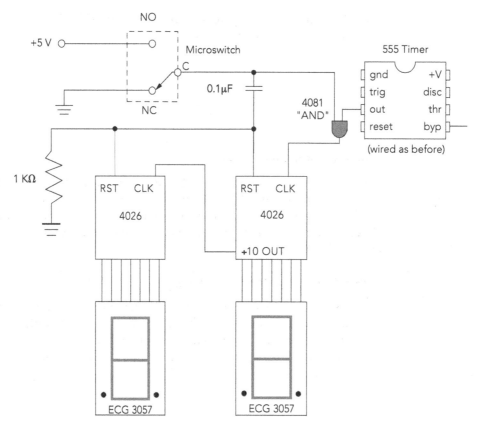

Fig. 25.14. A block wiring diagram for the digital stopwatch.

The output of the 555 Timer goes to one input on the AND IC, which acts as a gate, allowing the pulse train through only when the other input is held at 5 V (using the microswitch). The counters are reset by bringing the "reset" (RST) inputs on the counters to +5 V temporarily, also accomplished by the microswitch. (The additional capacitor/resistor combination between the microswitch and the reset terminals produces a brief positive-going pulse when the switch is first pressed. This positive pulse resets the counters to zero before they start counting.)

25.6.4. Activity: The Digital Stopwatch!

a. Use the circuit diagram of Fig. 25.14 to complete your stopwatch. Use your oscilloscope to adjust the 555 Timer to produce a pulse train with a frequency of 10 Hz (period of 0.1 s). This way the counter will count *tenths* of a second in the right-most digit and seconds in the left-most digit.

    b. Verify your stopwatch is working by depressing the microswitch for a short amount of time. According to your stopwatch, what is the shortest time you are able to close the microswitch for? What is the longest time you can depress the microswitch and still get an accurate time measurement with the stopwatch?

    c. When you have your digital stopwatch working, show it to your instructor or TA and get a signature witnessing that the stopwatch functions correctly!

[Signature] _____ [Date] _____

    d. Briefly explain what each part of the circuit accomplishes and how they work together to make the stopwatch.

UNIT 26: MAGNETS AND MAGNETIC FIELDS

This picture shows a small permanent magnet levitating above a high-temperature supercon-ductor. Although this phenomenon seems like magic, the levitation is due to an interaction between the magnet and induced currents flowing in the superconductor. But what exactly is magnetism, and what is the nature of the forces between magnets and moving charges? These questions (and many others) will be explored in this unit.

UNIT 26: MAGNETS AND MAGNETIC FIELDS

OBJECTIVES

1. To learn about the properties of permanent magnets and the forces they exert on each other based on the microscopic magnetic model.

2. To discover how a magnetic field produces a force on a moving charge, and how this leads to magnetic forces on current-carrying wires.

3. To understand why a charged particle moving perpendicular to a uniform magnetic field travels in circular motion, and how we can use this to measure the ratio of the electron charge to its mass, or e/m.

26.1 OVERVIEW

As a child you may have played with small magnets and compasses. Magnets exert forces on each other, and the small magnet that comprises a compass needle is affected by Earth's magnetism. Magnets are used in many common electrical devices, including meters, motors, and loudspeakers. Magnetic materials are essential in some types of computer hard drives, and large electromagnets made of current-carrying wires and iron are used to pick up whole automobiles or other materials in junkyards (see Fig. 26.1).

worradirek/Shutterstock

Fig. 26.1. A large electromagnet used to pick up scrap metal.

From a theoretical perspective, magnetism turns out to be an aspect of electricity rather than something separate. In the next two units we will explore the relationship between magnetic forces and electrical phenomena. For example, we'll see that permanent magnets can exert forces on current-carrying wires and vice versa. Electrical currents can even produce magnetic fields and changing magnetic fields can, in turn, produce electrical fields. In contrast to our earlier study of electrostatics, which focused on the forces between charges at rest, the study of magnetism is, at heart, the study of forces acting between *moving* charges.

MAGNETIC FORCES AND FIELDS

26.2 PERMANENT MAGNETS

The attraction of a magnet (e.g., a magnet on your refrigerator) is likely so familiar that you may not have thought much about how the attraction occurs. Let's begin our exploration of magnetism by carefully observing what occurs with permanent magnets and other objects. For the activities in this section, you will need:

- 2 rod-shaped permanent magnets (with ends marked)
- 1 aluminum rod (preferably the same size and shape as the magnets)
- A few paperclips
- 1 small piece of wood (e.g., wooden pencil)
- 1 small plastic object
- 5 small compasses
- 2 strings, 10 cm
- 1 piece Scotch tape, approx. 3″ long
- 1 table clamp
- 2 aluminum rods
- 1 right-angle clamp (aluminum preferable)

Interactions Involving Permanent Magnets

In the following activity, we explore the forces exerted by one magnet on another and then branch out to make qualitative observations of the forces that a magnet may exert on other objects. You should have a "bar magnet" (or "cylindrical magnet") at your table, one end of which is called the *north pole* (typically labeled "N" and/or colored red), while the other is called the *south pole* (labeled "S" and/or colored blue).

26.2.1. Activity: Permanent Magnets and Forces

a. What do you expect to happen when you bring two permanent magnets close to each other? Test out your predictions using two permanent magnets. Do the like poles attract or repel each other? What about the unlike poles? How do the rules of attraction and repulsion for magnets compare to those for electrical charges?

b. Consider four different objects: an aluminum rod, steel (e.g., a paper-clip), wood, and plastic. *Predict* what you think will happen when you bring each object near both poles of a magnet. For example, will it attract, repel, or neither?

Object	N Pole	S Pole
Aluminum rod		
Steel (e.g., paperclip)		
Wood (e.g., pencil)		
Plastic		

c. Now *observe* what happens when you bring the various objects close to each pole of the magnet and summarize your results in the table below.

Object	N Pole	S Pole
Aluminum rod		
Steel (e.g., paperclip)		
Wood (e.g., pencil)		
Plastic		

d. How do your predictions and observations compare?

You should have seen that magnets interact differently with different types of materials. When a magnet is brought close to another magnet, one can observe both attraction and repulsion, depending on whether like or unlike poles are in proximity. On the other hand, materials such as steel are attracted to both poles of a magnet, while some materials show no interaction at all. This leads us to classify three different types of substances in terms of their magnetic interactions: (1) permanent magnets, (2) magnetic materials (e.g., steel), and (3) non-magnetic materials (e.g., wood, plastic). Next, we'll consider what happens if a magnet is suspended in space.

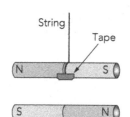

a. Antiparallel orientation

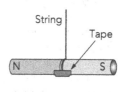

b. Parallel orientation

Fig. 26.2. Possible orientations for a hanging magnet suspended near an identical fixed magnet: (a) antiparallel orientation and (b) parallel orientation.

Magnet Orientations

Let's explore how a suspended magnet orients itself when it is placed close to another identical magnet, as well as what happens when the suspended magnet is placed far away from any other magnet. **Note:** If the experiment is not already set-up, you should tie the string *tightly* around the center of a magnet and put a small piece of Scotch tape under the string as shown in Fig. 26.2. Any rods and clamps used to support the magnet should be non-magnetic if at all possible.

26.2.2. Activity: Magnet Orientation

a. Consider holding a second magnet underneath a suspended magnet as shown in Fig. 26.2. Based on your previous observations, *predict* whether the suspended magnet will align itself parallel or anti-parallel to the magnet being held underneath. Briefly explain.

b. Try it out—suspend one magnet from a string and allow it to come to rest. Observe what happens when you *slowly* bring the second magnet up from below while maintaining it in a fixed orientation. How the does the suspended magnet behave? Which orientation does the suspended magnet adopt? Was your prediction correct?

c. Now move the second magnet far away and allow the suspended magnet to come to rest with no other magnets nearby. Look at your suspended magnet and compare it to the other suspended magnets in the room. Do the suspended magnets appear to orient in a particular direction (when no other magnets are nearby)? If so, what is the direction of the orientation, and why do you think it occurs?

d. Repeat the observation from part (c) with two compasses, being sure to keep the compasses away from each other and any other magnets. What do think a compass needle consists of? Briefly explain.

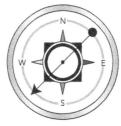

Fig. 26.3. A compass.

You should have observed that the suspended magnets and the compasses all align in the same orientation when no other magnets are present. Hopefully, you concluded that this is due to Earth, which itself behaves like a large bar magnet, having both a north pole and a south pole (see Fig. 26.4).

But notice that what we call the North Pole of Earth attracts the north pole of a magnet. Thus, Earth's *geographic* North Pole is actually a *magnetic* south pole. The reason for this confusing terminology is that before scientists understood magnetism, it was observed that one end of a magnet would point north and the other end would point south. The two ends of the magnet were therefore described as north (for north-pointing) and south (for south-pointing). Because the naming convention for magnets is the same as for geography, things can get a bit confusing![1]

Note: If you have a set of compasses, you may find that some of them point in the "wrong" direction. In other words, the north (red) pole of the compass needle may actually point south. This can occur when the bar magnet in the needle of the compass becomes magnetized the "wrong" way, suggesting that a magnet can change its north and south poles. To understand how this can happen and obtain a better understanding of how magnets work, we need to develop a conceptual model for magnetism.

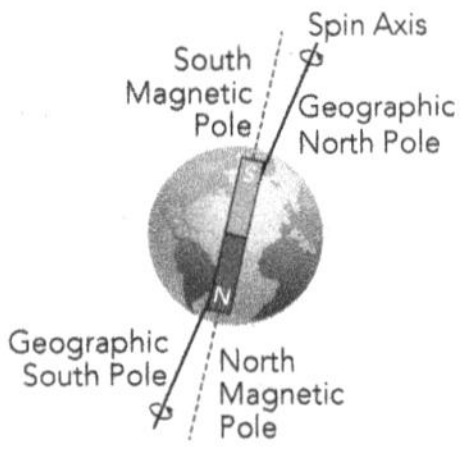

Fig. 26.4. The north pole of a magnet is attracted to the geographic North Pole (which is actually a magnetic south pole). Similarly, the south pole of the compass is attracted to the geographic South Pole.

26.3　THE MICROSCOPIC MAGNET MODEL

To motivate the microscopic magnet model, suppose you have several small disk magnets. Each one of these acts like a normal magnet with a north pole and a south pole. If you bring the north pole of one magnet close to the south pole of another, they will stick together. In fact, you can keep doing this, sticking three, four, or more of these disk magnets together (see Fig. 26.5). In the following activity we will consider how this stack of magnets compares to an individual disk magnet. For the activities in this section, you will need:

Fig. 26.5. An array of disk magnets: top set—stuck together, bottom set—pulled apart.

- 4 (or more) disk-shaped magnets
- 1 small compass or other magnetic field probe (e.g., Magnaprobe)
- 1 cylindrical or bar magnet

26.3.1. Activity: Magnetic Disks Together and Apart

a. Place the disk magnets in a stack and compare the behavior of the resulting stack with that of a similarly-sized, cylindrical magnet. Use a compass (or other magnetic field probe) to test whether the stack behaves like a regular cylindrical magnet. Does the stack of magnets appear to have a single north pole and a single south pole?

---

[1] It is probably worth a reminder that all of these terms—north pole, south pole, positive charge, negative charge—are *arbitrary* labels, so it doesn't really matter what they are called. At this point, we are stuck with the conventional names, even if they are sometimes confusing.

Fig. 26.6. Each piece of a broken magnet behaves as a regular magnet, with both a north pole and a south pole.

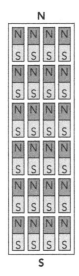

Fig. 26.7. A bar magnet can be thought of as being comprised of an array of miniature magnets all oriented the same direction.

b. You should observe that the stack of disk magnets behaves just like a cylindrical magnet. In fact, if you covered up the seams where the disks are stuck together, you probably could not tell the difference between the "stack magnet" and a normal magnet. Next, pull the stack apart into two stacks with two disks in each stack. Once again, compare the behavior of one of the resulting stacks with that of the cylindrical magnet. Does it also behave like a regular magnet with a single north pole and a single south pole?

c. It seems that no matter how we arrange the disk magnets, the stack always act like a single, regular magnet with a single north pole and a single south pole. It will be the same whether we have a single disk magnet, a stack of four, or a stack of twenty. So, here's a question: what do you think would happen if we cut one of the individual disk magnets in half? What would a split-magnet look like, based on our previous observation?

Although we can't do the experiment, you might guess based on what we have seen that the split-magnet would also act like a normal magnet, with a single north pole and a single south pole. This is, in fact, what happens—splitting a single disk magnet results in two new magnets, each with a single north pole and a single south pole. If we continue splitting the magnet into smaller and smaller pieces, we would see that we always end up with a new magnet having both a north pole and a south pole. In fact, a cylindrical magnet may get dropped and break into multiple pieces. When this occurs, it's easy to verify that each piece acts as an individual magnet (see Fig. 26.6). To understand why this is true, we need to look at the microscopic structure of a magnet.

Microscopic Magnets and Magnetic Materials

As discussed, breaking up a large magnet into smaller pieces results in entirely new magnets having both north and south poles. Thus, we can think of a big magnet as being comprised of a large number of miniature magnets all oriented in the same direction (see Fig. 26.7). Notice that the top of the big magnet consists of a large number of north poles from the mini magnets. If you were to bring a compass near this end, the south pole of the compass would be attracted to this set of north poles.

Together, all these mini north poles combine to make the north pole of the big magnet, with something similar occurring at the south pole. But notice that anywhere *inside* the magnet there is always a south pole right next to a north pole, effectively canceling each other out. For the big magnet, the poles only exist at the end where there are "unmatched" mini poles.

This model nicely explains why breaking a magnet always results in two complete magnets. Cutting a big magnet at any location would separate a line of mini north poles from mini south poles. The two new pieces would have an unmatched set of poles at new ends, and so each piece would have both a north pole and a south pole. No matter where you cut, you always end up with both a north pole and a south pole when you break off a smaller piece of a magnet.

It turns out this behavior continues all the way down to the microscopic level. Ultimately, magnetism arises primarily from the *magnetic moment* of the electron, which is due to the quantum-mechanical properties of the electron (typically its *spin*). We won't go into the details, but simply note that some elements (e.g., iron, cobalt, and nickel) have electron configurations that result in the atoms acting like microscopic mini-magnets, often referred to as *magnetic dipoles*. You can think of each one of these magnetic dipoles as a tiny arrow that resembles the magnet in a compass. The picture of a macroscopic magnet is then a large array of these microscopic magnets all lined up as shown in Fig. 26.8. This is the essence of the *microscopic magnet model*.

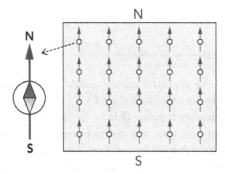

Fig. 26.8. Left: A microscopic magnetic moment due to an electron. Right: An array of magnetic moments all aligned inside a bar magnet. (Not drawn to scale!)

But what about other types of materials? For example, in Activity 26.2.1 we saw there are also magnetic materials that are not magnets on their own (a paperclip), as well as completely non-magnetic materials (aluminum, plastic, wood). Can the microscopic magnet model account for these types of materials as well?

26.3.2. Activity: Magnetic and Non-Magnetic Materials

a. The metal used in a magnetic material such as a paper clip typically contains some iron, which means that inside the material there are lots of these microscopic magnetic dipoles. But a paperclip on its own is not a magnet. In the space below, draw a rectangle representing the paperclip and sketch a number of mini magnetic dipoles. How do you think the dipoles must be oriented if the paper clip is *not* a magnet on its own? Briefly explain why you drew the dipoles the way you did.[2]

b. Now consider bringing the north pole of a permanent magnet up close to one end of your "paperclip rectangle." Remembering that each one of the magnetic dipoles in the paperclip acts like a little compass, draw a diagram showing what happens to the magnetic dipoles in the paperclip when the permanent magnet is nearby (or even touching). Does this explain why the paperclip is attracted to the permanent magnet?

c. Because the magnetic dipoles in the paperclip align due to the presence of the permanent magnet, you should see that the paperclip itself now acts a little bit like a magnet. Do you think it's possible to pick up a second paperclip using a paperclip that is in contact with a magnet? Try it out and write your observations below.

d. Next, hold the upper paperclip (the one touching the magnet) with your fingers and gently remove it from the magnet. Does the lower paperclip remain attached? What does this tell you about the upper paperclip?

e. Now flip the magnet over and slowly bring the *opposite* pole down toward the top of the upper paperclip (without touching it). What do you observe? Use the microscopic magnet model to explain what you see.

---

[2] Typically, it is not individual magnetic dipoles that are randomly aligned, but instead magnetic *domains*. Inside each domain the microscopic dipoles are aligned, but there are many of these domains with many different orientations resulting no net, macroscopic magnetic moment.

f. Finally, consider a non-magnetic material such as aluminum or plastic. The atoms in these materials tend to have electrons that come in pairs. That is, the microscopic dipole moments come in sets of two, and the two moments point in opposite directions.[3] In the space below, draw a rectangle representing a non-magnetic material and put in several magnetic dipole *pairs*. Explain why we do not see any magnetic interaction with this type of material.

In the previous activity, we saw how the microscopic magnet model can be used to explain magnets as well as the behavior of both magnetic and non-magnetic materials. In particular, a magnetic material such as a paperclip can become temporarily magnetized due to the presence of a permanent magnet. (This is somewhat similar to how materials can become electrically polarized due to the presence of excess charges nearby.) With this basic understanding of magnets, we are ready to proceed to the topic of magnetic fields.

26.4 MAGNETIC FIELDS

In electrostatics we began by talking about charges and electric forces before moving on to define the electric field. As you may recall, the electric field has both a magnitude and direction and can be represented using either vectors or field lines. In this unit we began by talking about magnets and magnetic forces,[4] so you may not be surprised to discover that our next topic is the *magnetic field*.

Fig. 26.9. Magnetic field lines around a "bar magnet."

When dealing with charges, convention dictates that the electric field points away from positive charges and toward negative charges. When dealing with magnets, we begin with an operational definition for the direction of the magnetic field: *the magnetic field direction is defined as the direction in which the north pole of a compass needle is oriented in the presence of the field*. The difference compared to the electric case is due to the fact that magnetic poles always come in pairs.

In the next activity, we will explore the alignments of small compasses at various places near larger magnets. This procedure will allow us determine the shape of the magnetic field from a magnet. For the observations that follow, you will need:

- 1 bar-shaped magnet
- 1 horseshoe magnet (optional)

---

[3] For example, the two electrons in the pair could be "spin-up" and "spin-down," resulting in no net magnetic moment for the pair.

[4] You may have noticed that when talking about forces between magnets and magnetic poles, we did not quantify the force as we did with Coulomb's law for the force between charges. Because magnetic poles always come in pairs (a magnetic dipole), the mathematics describing the interaction force between two magnets turns out to be much more complicated than the electric force between two charges.

- Several small compasses or other magnetic field probe (e.g., Magnaprobe)
- Iron filings
- Paper or other material for use with iron filings

26.4.1. Activity: Field Directions Around a Bar Magnet

a. Use a small compass or Magnaprobe to map out the magnetic field surrounding a large bar magnet. Sketch the magnetic field lines in the space below using arrows to denote the direction of the field based on the definition given in the paragraphs above. Remember that the strength of a field is represented by the density of the lines.

b. Lay a piece of paper over the bar magnet and sprinkle iron filings on it. It's helpful to place stacks of paper next to the magnet on each side so that the piece of paper on top of the magnet lies flat. Each little iron filing is like a compass needle that becomes magnetized and then aligns with the field (like a small magnetic dipole). In the space surrounding the magnet, do the iron filings give a similar picture as to the shape of the field around the magnet? Can the iron filings show the *direction* of the magnetic field?

c. Next, consider the region "inside" the magnet, or the region *between* two adjacent microscopic magnets inside the material. While you obviously can't place a compass inside the magnet, some of the iron filings in part (b) should be lying directly on top of the magnet; these should provide a hint about what the field looks like *inside* the magnet.

On the diagram below, redraw the magnetic field lines both inside and outside the magnet (including direction). You can assume the field lines are continuous across the boundaries between the magnet and the surrounding air. You will need to think carefully about the microscopic magnet model when determining which direction to draw the lines *inside* the magnet. (If you're having trouble, ask for help.)

While it can be tempting to draw the field lines inside the magnet above as pointing from left to right, the microscopic magnet model suggests otherwise. We can think about the inside of the magnet being comprised of rows of microscopic magnets, all aligned with their north poles pointing to the left in the diagram above. By definition, the north pole of these microscopic magnets point in the direction of the magnetic field, so the field actually points from right to left inside the magnet above. The result is that magnetic field lines form closed loops that go through the interior of the magnet, exit the north pole, circle around outside the magnet, and enter the south pole (see Fig. 26.10).

26.4.2. Activity: Magnetic Flux for a Bar Magnet

a. Let's return to think about the concept of flux, this time pertaining to the magnetic field (*magnetic flux*). For simplicity, we consider the two-dimensional situation of the field lines shown in Fig. 26.10. Draw several closed loops on this figure using dashed lines to represent closed Gaussian surfaces (in two dimensions). Let one loop enclose no magnetic pole, another loop enclose just one of one of the poles, and a third loop enclose both poles. Assuming that each field line coming into a loop is negative and each line coming out is positive, what is the net number of magnetic field lines coming out of a loop in each case?

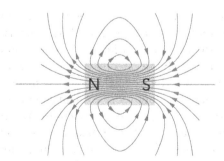

Fig. 26.10. A bar magnet showing the magnetic field both inside and outside the magnet.

Now we come to the magnetic equivalent of Gauss's law describing the net magnetic flux. As a reminder, Gauss's law is given by Eq. (20.4)

$$\Phi_{\text{elec}}^{\text{net}} = \oint \vec{E} \cdot d\vec{A} = \frac{q_{\text{encl}}}{\varepsilon_0} \quad \text{(Gauss's Law)}$$

where $\Phi_{\text{elec}}^{\text{net}}$ is the net electric flux through the closed surface, $\vec{E}$ is the electric field on the surface, $d\vec{A}$ is the area element vector, q_{encl} is the net charge enclosed by the surface, and ε_0 is a physical constant.

b. Based on your answer to part (a), can you predict what the net *magnetic* flux through any closed surface will be equal to?

c. Although clearly not a proof, part (a) suggests that the net magnetic flux through any closed surface will always be zero. Let's assume that to be the case. Recall that Gauss's law says the flux is proportional to the net charge enclosed by the Gaussian surface. If the magnetic flux is equal to zero in *all situations*, is it possible to have a net "magnetic charge" contained inside the surface? Briefly explain how your answer is consistent with the fact that splitting up a magnet always results in two new magnets, each having both a north and a south pole.

Activity 26.4.2 suggests that one will *never* find an isolated magnetic pole. In other words, magnetic charges don't exist. Just like we saw when breaking apart a magnet, you always end up with objects that have both a north pole and a south pole—magnetic poles always seem to come in pairs. This is what is observed for ordinary matter: there do not appear to be any magnetic "monopoles" like there are for electric charges.

26.4.3. Activity: The Magnetic Field from a "Horseshoe" Magnet

a. Consider bending a bar magnet so that it takes the shape of a U (often referred to as a "horseshoe" magnet). Predict what the magnetic field looks like in the region between the poles (in the region below N and above S). Sketch your predicted field lines.

b. Place a small compass or Magnaprobe between the poles of a U-shaped magnet and sketch the actual field line directions in the diagram below. Does this make sense based on the rules we developed for the bar magnet?

Armed with a basic understanding of permanent magnets and magnetic field lines, we now turn our attention to interactions between magnets and electric charges.

MAGNETIC FIELDS AND ELECTRIC CHARGES

26.5 CAN A MAGNET EXERT FORCES ON ELECTRIC CHARGES?

We know that a magnet exerts a force on another magnet or a magnetic material. We have also hinted that magnetic forces and electrical forces are related in some way. To start investigating this possibility, let's consider whether a magnetic field can exert a force on *electric* charges. To do the observations in this section, you will need:

- 2 pieces Scotch tape, 5–10 cm
- 1 bar or cylindrical magnet (with labeled poles)
- 1 aluminum post (non-magnetic)
- 1 electron beam demonstration device (e.g., open-frame demonstration oscilloscope, analog oscilloscope, *e/m* apparatus, etc.)

Magnetic Forces on Static Charges

Let's start by investigating possible forces between magnetic fields and *static* (stationary) electric charges. As we did previously, we'll use Scotch tape to create an electrically-charged object.

26.5.1. Activity: Is There a Magnetic Force on Static Charges?

a. Stick a piece of tape to the table and pull it up quickly (making it electrically charged). Now slowly bring the tape close to each pole of the magnet. Do you observe any interaction? Summarize your findings.

b. Next, do the same test with a *non-magnetic* metal bar or rod (e.g., aluminum). Bring the tape close to both ends of the non-magnetic bar. Do you observe any interaction? Summarize your findings.

c. Did you find any *difference* between the interactions of the tape with the magnet and the non-magnet? Based on these results, can you conclude that there must be a *magnetic* force on a static electric charge? Briefly describe why or why not. If not, what explains the interaction you observed in parts (a) and (b)?

You should have seen that the charged tape was attracted to the north pole of the magnet, the south pole of the magnet, and both ends of the non-magnetic bar. Because the interactions are all the same, it should be clear that the charged tape behaves the same way whether or not a magnet is present. In other words, there is no evidence for a magnetic interaction between a magnet and a stationary electric charge.

Magnetic Forces on Moving Charges

Let's now examine whether a magnet can exert a force on electric charges that are *moving*. We need a source of moving charges that can be visualized, and while we can't see individual electrons, there are methods for observing the path electrons follow.

As we saw in Unit 25, an analog oscilloscope directs an electron beam toward a screen; when the electrons hit the screen it glows, showing you the location of the beam. Alternatively, an apparatus commonly used for measuring the charge-to-mass ratio of the electron can also be used to visualize the path of an electron beam.

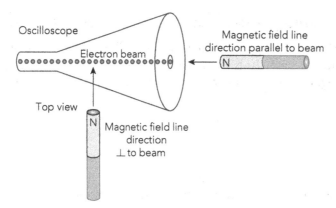

Fig. 26.11. Electron beam used to test whether a magnetic field can exert a force on a moving electric charge.

26.5.2. Activity: Observing Magnetic Forces on Moving Charges

a. Turn on the electron beam so that you can see either the full path of the beam or the location where the beam intersects the screen. Slowly move the north pole of your magnet closer to the beam in a direction *perpendicular* to the electron beam (see Fig. 26.11). Does the electron beam experience a force? Based on how it moves, what is the direction of the force? Sketch vectors showing the direction of the magnetic field (based on the bar magnet), the direction of motion of the electron beam (before it is deflected), and the direction of the force on the beam.

b. Now, slowly move the north pole of your magnet closer to the beam in a direction *parallel* (or *antiparallel*) to the electron beam (see Fig. 26.11). Does the beam appear to experience a significant force now?

You should have found that when the magnetic field is parallel to the direction of motion of the charges, there is essentially no force on the charges. However, when the magnetic field is perpendicular to the motion of the charges, there is a force that's perpendicular to *both* the direction of the magnetic field and the direction of motion of the charges. You might recall that we have seen situations in the past where two vectors are multiplied together to produce a third vector that's perpendicular to both of the original vectors; this relationship arises from the *vector cross product*.

We briefly review the vector cross product here; if you wish, you can return to Section 13.6 for a full review. The cross product between two arbitrary vectors $\vec{A}$ and $\vec{B}$ is defined to be $\vec{A} \times \vec{B} = AB \sin \theta \, \hat{n}$, where θ represents the *smaller* angle between the directions specified by the vectors. The magnitude of the cross product is given by $\left| \vec{A} \times \vec{B} \right| = AB \sin \theta$, while the direction is given by $\hat{n}$, which is perpendicular to both $\vec{A}$ and $\vec{B}$ and determined by the right-hand rule.

The Lorentz Force

The magnetic force on a moving charge can be calculated from the equation

$$\vec{F}_{\text{mag}} = q\vec{v} \times \vec{B} \tag{26.1}$$

where q is the charge (which can be positive or negative), $\vec{v}$ is the velocity of the charged particle, and $\vec{B}$ is the magnetic field vector. You may have noticed that, thus far, we have not discussed the strength of the magnetic field. One reason for avoiding this topic is because calculating the magnetic field is typically more complicated than for the electric field. But another reason is because Eq. (26.1) serves as a rather backwards mathematical definition of the magnetic field $\vec{B}$.

That is, the magnetic field $\vec{B}$ can be *defined* as that vector which, when crossed into the product of the charge and its velocity, leads to a force given by $\vec{F}_{\text{mag}} = q\vec{v} \times \vec{B}$. The standard unit for $\vec{B}$ that follows from Eq. (26.1) is newtons per coulomb-meter per second, and this unit of magnetic field is known as the *tesla* (abbreviated T).

In Unit 19, we saw that a charged particle (whether moving or not) experiences a force in an *electric* field: $\vec{F}_{\text{elec}} = q\vec{E}$. If a charged particle is in a region that contains both an electric and magnetic field, the total force is the sum of these two forces, typically referred to as the Lorentz force:

$$\vec{F}_{\text{Lorentz}} = q\vec{E} + q\vec{v} \times \vec{B} \tag{26.2}$$

26.5.3. Activity: Working with the Magnetic Force

a. Recall that the magnetic field outside a magnet points away from the north pole. Look back at your results for the perpendicular situation in Activity 26.5.2 and show that the vector cross product $\vec{F}_{\text{mag}} = q\vec{v} \times \vec{B}$ properly describes the relative directions of the three vectors involved. Be sure to draw a diagram showing all three vectors. **Hint**: Don't forget that q is negative in the case of an electron beam!

b. Given the orientation of the vectors, what is the angle between $\vec{v}$ and $\vec{B}$ (when there is no deflection)? Using Eq. (26.1), determine the magnitude of the magnetic force for this situation in terms of q, v, and B.

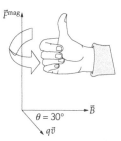

c. Now consider a different situation. Assume a proton is traveling at $\theta = 30$ degrees with respect to a magnetic field of strength 2.6×10^{-3} T, as shown in Fig. 26.12. If the proton is traveling at a speed of 3.0×10^6 m/s, what is the magnitude of the force exerted on the proton by the magnetic field? As usual, the direction of the force is given by the right-hand rule, as shown in Fig. 26.12.

Fig. 26.12. Right-hand rule to find the direction of a magnetic force exerted on a moving charge. The force vector is perpendicular to both $\vec{v}$ and $\vec{B}$.

d. If the particle in part (c) were an electron instead of a proton, what would be the magnitude and direction of the force exerted on the particle by the magnetic field? **Hint**: No calculations are needed here!

26.6 MAGNETIC FORCE ON A CURRENT-CARRYING WIRE

We just observed that magnetic fields can deflect a beam of electrons by exerting a force on the moving charges. This brings up an interesting question: if a current-carrying wire is placed near a magnet, will it experience a force? To examine this situation, you will need:

- 1 lantern battery, 6 V
- 2 alligator clip leads, 20 cm
- 2 alligator clip leads, 10 cm
- 1 normally-open switch
- 1 strong magnet, 7.5 kGauss or more
- Table clamps, right-angle clamps, and rods to hold wire (optional)

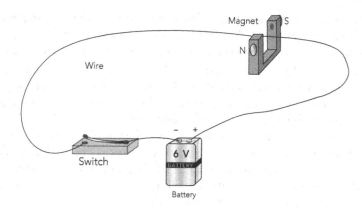

Fig. 26.13. When the switch is closed, the wire carries current in a magnetic field.

26.6.1. Activity: Magnetic Force on a Wire

a. Predict the direction of the force on a (non-magnetic) current-carrying wire when it is placed between the poles of a strong permanent magnet like in Fig. 26.13. Think carefully about the directions of the current and magnetic field.

b. Set up the circuit shown in Fig. 26.13 and place the wire between the poles of the magnet while your partner closes and opens the switch. (This experiment may be done as a demonstration.) **Warning**: The switch should only be closed for a short time or the battery will quickly die. Describe your observations, including the direction of the force on the wire. (If you wish, you can change the direction of the current by reversing the battery connections or the direction of the wire.)

c. Thinking about the relative directions of the current, the magnetic field, and the resulting force on the wire, verify that your observations are consistent with Eq. (26.1).

The previous activity demonstrates that a current-carrying wire experiences a force due to a magnetic field. A current represents a collection of many charged particles in motion. Assuming the wire is not parallel to the magnetic field (in which case $\vec{v} \times \vec{B} = 0$), moving charges in the wire experience a magnetic force in a direction perpendicular to the wire. Because the charges are constrained to remain in the wire, this results in a force on the wire itself (at least the portion of the wire inside the region of the magnetic field).

In principle, one could figure out the number of electrons moving in the wire and their average speed to determine this force. Practically speaking, it would be much easier to express the force using quantities that are more easily measured. In the following activity, we will work to rewrite the magnetic force equation in terms of the current flowing in the wire.

26.6.2. Activity: Relating the Magnetic Force to Current

a. Imagine you have a straight wire of length L carrying a current i (see Fig. 26.14).[5] If a single charge carrier of charge q (assumed to be positive) takes a time Δt to traverse the distance L, write down an expression for the *drift velocity* $\langle \vec{v} \rangle$, the average velocity of the charge carriers in the wire. **Note**: The vector $\vec{L}$ represents the displacement of the charges along the wire and points in the direction of (positive) current flow.

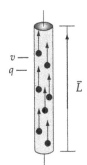

Fig. 26.14. The vector $\vec{L}$ represents the displacement of positive charge carriers moving at an average velocity $\langle \vec{v} \rangle$.

b. If a total amount of charge Q travels out the end of the wire in the time Δt, write down an expression for the current i in the wire in terms of Q and Δt.

c. The quantity Δt is present in both your expressions for parts (a) and (b). Solve these equations for Δt and set them equal to each other. Then solve the resulting equation for the product $Q \langle \vec{v} \rangle$.

---

[5] If the entire wire is in a region of magnetic field, L is the length of the wire; if only a portion of the wire is in the region with field, L is the length of this portion. (Portions of the wire that are not in a magnetic field experience no magnetic force).

d. You should have found that $Q\langle\vec{v}\rangle = i\vec{L}$. Equation (26.1) $\left(\vec{F}_{mag} = q\vec{v} \times \vec{B}\right)$ involves a similar product $q\vec{v}$, where $\vec{v}$ was the velocity of a single charge q. If we consider the collection of positive charges to be moving at the same average velocity, we can use the result of part (c) to write $\vec{F}_{mag}$ in terms of i, $\vec{L}$, and $\vec{B}$. Write down such an expression in the space below.

You should have found that the magnetic force on a current-carrying wire is given by

$$\vec{F}_{mag} = i\vec{L} \times \vec{B} \tag{26.3}$$

Note that Eq. (26.3) is simply a more convenient way of expressing Eq. (26.1) when dealing with macroscopic currents and wires. At base, they both say the same thing. Equation (26.3) can be generalized to the case when the wire is not straight. In this situation you would need to consider small wire segments $d\vec{L}$, each of which experiences a force $d\vec{F}_{mag}$. The total force $\vec{F}_{mag}$ on the wire would then be the vector sum (or integral) of all the $d\vec{F}_{mag}$ contributions.

26.7   MAGNETIC FORCE AND TORQUE ON A CURRENT LOOP

The combination of a current-carrying wire loop and permanent magnet lie at the heart of many important devices. To understand how these two objects interact, let's consider what happens when a rectangular loop of wire carries a current in the presence of a magnetic field.

Assume there is a current i flowing around the rectangular loop shown in Fig. 26.15. The current loop begins in the plane of the paper and is initially at

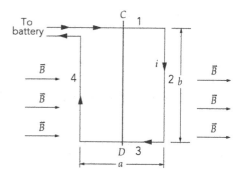

Fig. 26.15. A rectangular loop of wire with dimensions a and b in a magnetic field of magnitude B.

rest. A magnetic field parallel to the plane of the paper (and pointing to the right) is also present. The loop is free to rotate about the line *CD*.[6]

For your observations in this section, you will need:

- 1 rectangular wire loop (with multiple turns), R approx. 2 Ω
- 1 lantern battery, 6 V
- 3 alligator clip leads, approx. 20 cm
- 1 U-shaped magnet, approx. 2.5″ gap
- 1 SPST switch
- 1 small compass (optional)
- 1 ammeter (optional)
- Table clamps, right-angle clamps, and rods to hold loop (optional)
- Electric motor demonstration (optional)
      OPTIONAL (basic electric motor)
  - 1 1.5 V battery, alkaline
  - 1 1.5 V battery holder with copper alligator clip leads
  - 1 small, insulated copper wire loop, approx. 10 turns, with one end stripped all the way around and one end stripped only half-way around
  - 1 strong disk magnet
  - 2 paperclips
  - 1 rubber band for holding paperclips (or other method for suspending wire loop above magnet)

26.7.1. Activity: Predicted Force and Torque on a Current Loop

a. Use the magnetic force law of Eq. (26.3) and the right-hand rule to determine the predicted direction of the magnetic force on each segment of the wire in Fig. 26.15 (consider segments 1, 2, 3, and 4 separately). Describe the direction of the force on each segment using terms such as "to the right," "to the left," "up," "down," "into the page," and "out of the page." Remember that *i* is defined to be the flow of positive charge, and the vector $\vec{L}$ points in the direction of current flow along each segment (indicated by the arrows).

b. Is there a *net force* on the loop? Briefly explain.

---

[6] Although we analyze a single loop of wire, in principle there are typically many turns of wire around the same loop, each of which will experience the same forces (and torques).

c. Is there a *net torque* on the loop about the axis *CD*? Based on this, what motion of the loop do you predict will result (at least initially)? If you need a review of torque, see Unit 13.

d. Now suppose the initial torque results in the plane of the wire loop being rotated 90° about the axis through points C and D so it is now *perpendicular* to the plane of the paper (with side 2 above the page and side 4 below). Assuming the direction of the magnetic field is unchanged, is there a *net force* on the loop at this instant? Briefly explain.

e. Is there a *net torque* about the axis *CD* at this instant? Based on your answer, and the fact that the loop would have some rotational momentum as it reaches this point, what do you expect to happen to the loop as it reaches the 90° position?

f. Finally, what should happen to the forces and torques if the current in the loop increases? What about if the magnetic field increases?

You should have found that while there is no net force on the loop, there is initially a net torque that causes the loop to rotate about the axis *CD*. Although this torque diminishes as the loop rotates, the torque continues to twist the loop until the loop reaches the 90° position. At this instant, you should have found that both the net force and net torque (about the *CD* axis) are zero. However, because the loop is rotating as it reaches this position, its rotational momentum causes it to continue to rotate past the 90° position (even though the torque at that instant is zero).

Although we won't do so, one can use the same procedure to show that after the loop passes the 90° position, the net torque reverses direction, which acts to reduce the loop's rotational velocity. Thus, for the arrangement shown in Fig. 26.15, the 90° position is the stable resting location when the current and field are both present.

To observe the torque on a current loop, we need to place the loop in a magnetic field and pass an electric current through the wire. We will use the space between the poles of a U-shaped magnet for the magnetic field. We mapped out the magnetic field between the poles of such a magnet in Activity 26.4.3. For the required observations, you should wire a single 6 V battery, rectangular loop, and switch in series. Note that the torque will be larger if you use a wire with multiple loops on it (why?).

26.7.2. Activity: Observed Force and Torque on a Current Loop

a. Place the loop of wire so its plane is parallel to the magnetic field between the poles of the magnet (as depicted in Fig. 26.15). Briefly close the switch to pass a current through the loop. What happens? Does this agree with your prediction from Activity 26.7.1?

b. Next, hold the loop of wire so its plane is perpendicular to the magnetic field between the poles of the magnet (at the 90° position). Briefly close the switch and let go of the loop. What happens? How does this compare with your prediction?

c. Place the loop back in its original starting position. What happens if you reverse the wire connections to the battery so the current travels through the loop in the opposite direction?

d. As you might imagine, a current loop in a magnetic field has practical implications. One example is a device to measure current. Can you think of how you might use this device to measure the amount of current flowing in the wire? (Such device is known as a *galvanometer* and is used to measure current or voltage in analog meters.) **Hint**: You can use a spring to resist the rotation of the loop.

e. It may have occurred to you that we could make an extremely useful device if we could figure out a way to keep the loop rotating (instead of it rotating to the 90° position and stopping). Can you think of a way to keep the net torque always acting in the same direction? What would be required to maintain the same direction of the torque after the loop passes the 90° position? What might you call this device that uses an electric current to make something continuously rotate?

We just saw how a current-carrying loop of wire can experience a torque in a magnetic field. The magnitude of the torque is maximum when the plane of the loop is parallel to the magnetic field, and minimum (zero) when the plane of the loop is perpendicular to the magnetic field. In the following activity, we explore the maximum torque on the loop in more detail.

26.7.3. Activity: Maximum Torque on a Rectangular Loop

Return to Fig. 26.15 and your answers to Activity 26.7.1. Considering the case when the plane of the loop is *parallel* to the magnetic field, calculate the magnitude of the net torque on the loop about axis CD in terms of the current i and loop dimensions a and b.

You should have found that the net torque involves the product ab, which is the area enclosed by the loop. As we discuss in the following section, this turns out to be a general result.

Quantitative Torque on Current Loop

This process can be extended to the case of non-rectangular loops. When doing such calculations, one typically generalizes Eq. (26.3) to the case of small wire segments: $d\vec{F}_{\text{mag}} = i\,d\vec{L} \times \vec{B}$. The forces and torques on each segment can then be added together vectorially (or integrated) to find the net force or net torque about an axis. Regardless of the shape of the loop, it turns out that the maximum torque on a loop is given by $\tau = iAB$ (where A is the area of the loop).

As we will see, a small current loop produces a magnetic field that looks just like the field from a small disk magnet. This fact helps one conceptually understand the behavior of a current loop, and it's useful to define a *magnetic dipole moment* for a current loop

$$\vec{\mu}_{\text{loop}} = i\vec{A} \ \left(\text{dipole moment of a current loop}\right) \tag{26.4}$$

where A is the area of the loop. The direction of $\vec{A}$ is given by yet another right-hand rule: wrap your fingers in the direction of current around the loop, and your thumb gives the direction of $\vec{A}$ (and hence $\vec{\mu}$).

We can make use of our new definition in many calculations. For example, determining the torque on a current loop can be generalized for any orientation of the loop relative to the magnetic field, with the torque on an arbitrary loop given by

$$\vec{\tau} = \vec{\mu}_{\text{loop}} \times \vec{B} \quad (\text{torque on current loop}) \tag{26.5}$$

Note that when $\vec{\mu}_{\text{loop}}$ is parallel to $\vec{B}$, the cross-product results in zero; this is the stable position. This result can be easily understood by thinking of the current loop as a small disk magnet—the current loop orients itself so that it is aligned with the magnetic field.

An Electric Motor

In part (e) of Activity 26.7.2 we considered how a current loop in a magnetic field can create an *electric motor*, or a device that continually rotates when a current is applied. Ideally, one would want to reverse the current direction in the middle of a revolution so that the torque always acts in the same direction. Although this can be achieved without too much effort, a basic electric motor can be constructed by simply turning off the current for half a rotation, relying on the rotational momentum of the loop to complete a full revolution.

In the following activity, your instructor may ask you to construct a simple electric motor using a battery, a magnet, and small loop of wire (there may also be a demonstration electric motor). An example of such a motor is shown in Fig. 26.16.

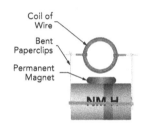

Fig. 26.16. A very simple electric motor.

26.7.4. Activity: A Simple Electric Motor

Using the equipment listed at the beginning of this section and Fig. 26.16 as a guide, see if you can construct a functioning electric motor. The goal is to have the wire loop continually spin when the current is flowing. Can you get the loop to spin in either direction? **Note:** You may need to give the wire loop a small "kick" to get it started.

THE *e/m* EXPERIMENT

26.8 A CHARGED PARTICLE IN A UNIFORM MAGNETIC FIELD

We have seen that a charged particle moving perpendicular to a magnetic field experiences a magnetic force that's perpendicular to both its velocity and the magnetic field. In the next activity, we'll consider the trajectory of such a particle. While interesting in its own right, the analysis also prepares us to perform a measurement on the electron known as the "e/m experiment." This experiment allows one to determine the charge-to-mass ratio, or e/m, for the electron. Back when physicists were trying to determine precise values for these quantities, it was difficult to establish exact values for an electron's charge or mass. Being able to reliably measure the ratio e/m helped to nail down the specific values for e and m.

You will need the following items for the activities in this section:

- 1 ball attached to a string, 1.5 m (optional)
- 1 e/m apparatus with Helmholtz coils
- 1 Magnaprobe or small compass

Path of an Electron in a Magnetic Field

Consider an electron that is shot with velocity $\vec{v}$ from left to right in the presence of a uniform magnetic field $\vec{B}$ directed into the paper (indicated by the ×'s in Fig. 26.17).

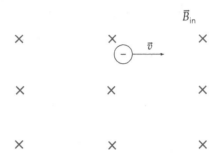

Fig. 26.17. An electron traveling in a uniform magnetic field perpendicular to its direction of motion.

26.8.1. Activity: Magnetic Force on an Electron

a. At the instant shown in Fig. 26.17, determine the direction of the magnetic force on the electron. Sketch a vector showing this force on the charge. **Hint**: Use Eq. (26.1) and don't forget that the electron has a negative charge.

b. A short time later, the electron will be traveling in a slightly different direction. Sketch in a new velocity vector in Fig. 26.17 that represents this, with the tail of your vector beginning at the trip of the original velocity vector.

c. Determine the direction of the magnetic force on the electron at this later time using your new velocity vector. How will this force cause the electron's velocity to change a moment later? Sketch yet another velocity vector in Fig. 26.17 that represents the electron's velocity at this later time.

d. The rough displacement of the electron traveling over the first few moments should look something like that shown in Fig. 26.18. Complete the diagram showing the rough path of the electron in the magnetic field as it continues moving.

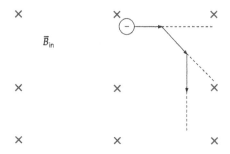

Fig. 26.18.

e. Suppose we broke the path in Fig. 26.18 into a very large number of very tiny steps. What would the approximate shape of the path be? How do you think the path would change if you increased or decreased the magnitude of the magnetic field? Briefly explain.

Hopefully, you convinced yourself that the path of an electron in a uniform magnetic field appears to be a circle.[7] We also know the magnetic force is always perpendicular to the electron's velocity, so the work done by the magnetic field is equal to zero: $W_{\text{mag}} = \int \vec{F}_{\text{mag}} \cdot d\vec{r} = 0$. Moreover, assuming the magnetic force is the only force acting, the electron's change in kinetic energy will also be zero: $W_{\text{net}} = \Delta K = 0$. The electron therefore moves in circular motion with a constant speed, or what we refer to as *uniform circular motion*.

---

[7] We are assuming an initial velocity directed perpendicular to the magnetic field. If the electron has some component of its velocity parallel to the magnetic field, it will still move in circles but will also move along the direction of the magnetic field, thereby tracing out a helix.

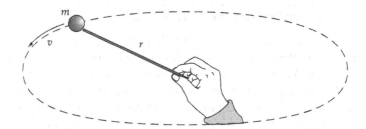

Fig. 26.19. Swinging a ball of mass m in a circle of radius r at constant speed v.

26.8.2. Activity: Review of Uniform Circular Motion

a. An example of uniform circular motion, which we studied in Unit 7, is a ball being swung around on a string with a constant speed (see Fig. 26.19). If the ball is being swung quickly, the effect of gravity can be ignored. In such a situation, what force is responsible for keeping the ball moving in circular motion? In what direction does this force act?

b. Is the ball accelerating as it moves? If so, in what direction is the acceleration?

c. Use Newton's second law to write down an equation relating the magnitude of the tension force F_T in the string, the mass m of the ball, the speed v of the ball, and the radius r of the circle. If you don't remember how to do this, you should review the treatment of centripetal acceleration in Unit 7.

To measure e/m for an electron, we will use an electron gun with a potential difference ΔV to accelerate a beam of electrons to a speed v. The electron beam then enters a region of uniform magnetic field B while moving perpendicular to the field (just like in Fig. 26.17). The electrons will move in a circle of radius r, and we want to find a theoretical equation that predicts the value of e/m using measurements of ΔV, B, and r. The following activity walks through this analysis.

26.8.3. Activity: Deriving the *e/m* Equation

a. We start with the electron gun (where there is no magnetic field). In this region the electrons are accelerated to a final speed v by a potential difference ΔV. Assuming the electrons start at rest, use conservation of mechanical energy to find an expression for the ratio e/m in terms of the final speed v and the potential difference ΔV. **Hint**: You might find it easiest to forget about signs and consider only magnitudes.

b. Next, the electrons enter the region of magnetic field with speed v, where the magnetic force in Eq. (26.1) causes the electrons to move in a circle of radius r. Use this fact, along with Newton's second law, to find an expression for the ratio e/m in terms of the final speed v, the radius r, and the magnetic field B. There is only a single force here, so you can simply equate magnitudes.

c. You should now have two expressions for the ratio e/m, both of which involve the speed v. This speed is challenging to measure experimentally (the electrons are very small and moving very fast!), so we would like to find an expression that involves only the radius r, the magnetic field B, and the potential difference ΔV (all of which are easily measurable). Use these two expressions to eliminate v and find an equation for e/m only in terms of r, B, and ΔV. There are multiple ways to do this—whatever you find easiest is fine!

You should have found that the charge-to-mass ratio for the electron is given by

$$\frac{e}{m} = \frac{2\Delta V}{r^2 B^2} \qquad (e/m \text{ ratio}) \qquad (26.6)$$

where ΔV is the magnitude of the potential difference applied to the electron gun, B is the magnitude of the magnetic field in the region, and r is the radius of the electron's orbit. Armed with this equation, we are now in a position to perform an experiment to obtain values for r, B, and ΔV, from which we can determine e/m for an electron!

Generating and Calculating the Magnetic Field

We need to create a uniform magnetic field with a known strength. Up until now, we have been using bar magnets (or horseshoe magnets) for our magnetic fields. But the fields from these types of magnets are far from uniform—they change direction rapidly and get weaker as you move away from the ends.

Instead of using permanent magnets, the typical e/m apparatus uses an *electromagnet*. In an electromagnet, the magnetic field is created by running a current through some coils of wire (we'll be studying the ability of currents to produce magnetic fields in the next unit!). One can create a reasonably uniform magnetic field using two large, current-carrying coils called *Helmholtz coils*, shown in Fig. 26.20.

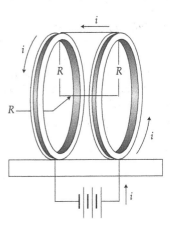

Fig. 26.20. Helmholtz coils are used to produce a uniform magnetic field. The distance between the coils is equal to the radius of the coils.

In the next unit, we will learn more about how currents in wires produce magnetic fields. For now, we simply use the fact that for a pair of Helmholtz coils of radius (and separation distance) R, the magnitude of the magnetic field in the region near the center of the arrangement is given by

$$B = \frac{8\mu_0 Ni}{R\sqrt{125}} \quad \text{(magnetic field from Helmholtz coils)} \qquad (26.7)$$

where

μ_0 is the magnetic permeability (a constant equal to 1.26×10^{-6} Tm/A)

N is the number of turns in each coil

i is the current through the coils in amps

R is the radius of the coils (and the distance between them) in meters

Your instructor will provide you with the details of your particular e/m apparatus and may demonstrate the experiment to the class.

26.8.4. Activity: The Magnetic Field Inside the Coils

a. Record the specifics of your Helmholtz coils in the table below. Numbers for a couple common e/m apparatuses are given at the bottom for reference.

Measurement	Symbol	Value	Units
Helmholtz coil current	i		amperes
Number of turns of coil wire[1,2]	N		—
Radius of Helmholtz coil[1,2]	R		meters
1. For a standard Welch e/m apparatus: $N = 72$ and $R = 0.33$ m. Typical filament to pin distances (that is, $2r$) are: pin #1 (6.48 cm); pin #2 (7.75 cm); pin #3 (9.02 cm); pin #4 (10.3 cm); pin #5 (11.5 cm). 2. For the Pasco SE-9625 apparatus: $N = 130$ and $R = 0.15$ m.			

b. Turn on the recommended current through the coils. Hold a Magnaprobe or small compass between the coils as close to the center of the arrangement as possible (this can be challenging due to the glass bulb). Does the magnetic field appear to be perpendicular to the plane of the coils?

c. Use Eq. (26.7) and the values of your parameters to compute the magnitude of the magnetic field between the coils.

Determining *e/m*

Now that we know the magnetic field strength, we are ready to calculate a value of *e/m* using our experimental measurements. A detailed view of a Welch *e/m* apparatus is shown in Fig. 26.21. While your set-up may look a little different, the basic idea is the same. We know the magnitude of the magnetic field *B* from Activity 26.8.4, so we must determine the accelerating potential ΔV and radius *r* of the orbit. For a given accelerating potential, the electron beam will trace out a circle of some radius. By varying the accelerating potential and recording the corresponding radii, we will have a series of data points from which to calculate the ratio *e/m*.

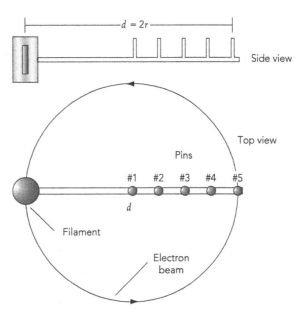

Fig. 26.21. Views of the inside of a Welch *e/m* apparatus. Your apparatus may look a bit different (and may not have pins).

26.8.5. Activity: Determining *e/m* Experimentally

a. Record a series of measurements using at least five different values of ΔV and *r*. Enter these measurements into a spreadsheet and use Eq. (26.6), along with your value of *B*, to calculate the ratio *e/m* for each measurement.[8] Then determine the average, the standard deviation, and the standard deviation of the mean for your set of measurements. Write down (or print out) your results in the space below. Be sure to include proper units!

---

[8] Alternatively, your instructor may have you vary the magnetic field *B*. In any case, you want a series of measurements for determining *e/m*.

b. Look up the accepted values of e and m and calculate the ratio e/m. Write this result in the space below.

c. How close is your measured (average) ratio to the accepted value? If your average is off by more than one or two standard deviations of the mean, you probably have a systematic error. If this is the case, do you best to account for the systematic error.[9]

---

[9] Depending on the details of your set-up, there may be an additional effect that needs to be considered (can you think of what it is?). Your instructor may provide directions to account for possible systematic error.

26.9 PROBLEM SOLVING

Now that we have analyzed the e/m experiment, we are ready to see a real-world application of a very similar arrangement. Figure 26.22 shows a *mass spectrometer*, which can be used to identify molecules in a sample by their charge-to-mass ratio q/m (you may have used such a device in a chemistry lab). Molecules in the sample are ionized by removing one (or more) electrons, and the positively-charged ions are accelerated from rest through a potential difference ΔV. The particles then enter a region of uniform magnetic field directed into the page where they undergo circular motion due to the magnetic force. After tracing out half a circle, the molecules hit a spatial detector that registers where the particles arrive, from which the radii of motion can be determined.

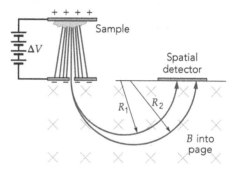

Fig. 26.22. Diagram showing a basic mass spectrometer.

Although all the ions have the same charge (assuming single ionization), different molecules will, in general, have different masses. This means that the various molecules comprising the sample end up at different positions on the detector, allowing one to determine the composition of the sample.[10]

26.9.1. Activity: The Mass Spectrometer

Assume the magnetic field has a magnitude of $B = 0.200$ T and that the accelerating potential is $\Delta V = 100$ V. If the sample is just air, the signal consists primarily of diatomic nitrogen $\left(N_2^+\right)$ and diatomic oxygen $\left(O_2^+\right)$. Find the radii R_1 $\left(N_2^+\right)$ and R_2 $\left(O_2^+\right)$ of the particle trajectories. **Hint**: The atomic mass unit has a value of $u = 1.66 \times 10^{-27}$ kg.

---

[10] Technically, the device only allows one to distinguish unique charge-to-mass ratios.

UNIT 27: ELECTRICITY AND MAGNETISM

The image above shows a light bulb attached to a coil of wire. The bulb is glowing, even though the coil is not connected to anything else—it's simply being held in the air above another device! How is this feat accomplished? In this unit you will learn about how currents produce magnetic fields and how changing magnetic fields in turn produce currents. The wireless charging station you may use to charge your phone is based on this process.

UNIT 27: ELECTRICITY AND MAGNETISM

OBJECTIVES

1. To develop an understanding of the magnetic field produced by a current-carrying wire and to use Ampère's law to calculate the magnetic field for a long, straight wire.

2. To observe that an electric field can be produced by a changing magnetic field through a process known as *induction*.

3. To explore the mathematical properties of induction as expressed in Faraday's law and to verify Faraday's law experimentally.

27.1 OVERVIEW

In the previous unit we observed that permanent magnets exert forces on both freely moving charges and electric currents in conductors. We described such interactions by referencing the magnetic field of the magnet. Newton's third law states that whenever one object exerts a force on another, the latter object exerts an equal and opposite force on the former. Thus, if a magnet exerts a force on a current-carrying wire, does that mean the wire exerts an equal and opposite force back on the magnet? If so, how does this happen?

It seems plausible that the symmetry demanded by Newton's third law should lead us to hypothesize that if moving charges feel a force as they pass through a magnetic field, then these charges should be capable of exerting forces on the sources of the magnetic field. In fact, it is not unreasonable to speculate that currents and moving charges exert these forces by *producing magnetic fields themselves.*

This line of reasoning can lead to further speculation. Because charges also have electric fields associated with them, moving charges can produce changing electric fields. Might these changing electric fields be the cause of the magnetic fields? If so, then, by symmetry, *can changing magnetic fields produce electric fields?*

This unit deals with two primary questions, both related to the interaction of electric and magnetic phenomena. First, do moving charges (like those in a current-carrying wire) produce magnetic fields? And second, does a changing magnetic field produce electric fields (and possibly currents)? Investigating these questions will lead us to Ampère's law (relating magnetic fields and currents) and Faraday's law (relating changing magnetic flux to electric fields and currents).

Faraday's law lies at the heart of electricity and magnetism and is one of the most profound laws in classical physics. Combining Ampère's law with Faraday's law allows us to describe how electricity produces magnetism and how magnetism produces electricity. Thus, electricity and magnetism are ultimately different aspects of the same phenomenon. At the end of this unit we will reformulate the laws of electricity and magnetism into a set of four interdependent expressions known as Maxwell's equations.

MAGNETISM FROM ELECTRICITY

27.2 THE MAGNETIC FIELD NEAR A CURRENT-CARRYING WIRE

In 1819 the Danish physicist Hans C. Oersted placed a current-carrying wire near a compass needle during a lecture demonstration before a group of students. In the activities that follow, we will investigate the characteristics of the magnetic field produced by currents. To do the activities in this section, you will need:

- 1 lantern battery, 6 V
- 1 D-cell battery, 1.5 V
- 1 D-cell battery holder
- 3 alligator clip leads, 20 cm
- 1 normally-open switch
- 1 rod stand
- 1 aluminum three-finger clamp (non-magnetic)
- 2 aluminum support rods (non-magnetic)
- 1 right-angle clamp
- 1 mass or tape (to hold down wire)
- 1 plexiglass (or cardboard) sheet, approximately 8″ × 8″ with hole cut in middle
- 6 small compasses

27.2.1. Activity: Predictions for Magnetic Fields from Currents

a. Assume that we will see a magnetic field in the vicinity of a straight, current-carrying wire. Do you expect the magnitude of the field to depend on the distance from the wire? If so, will it increase, decrease, or stay the same as the distance from the wire increases? Briefly explain.

b. What do you think will happen to the magnitude of the magnetic field if the current is increased or decreased? Briefly explain.

c. Here's a tougher question. In which *direction* do you think the magnetic field will point? And what do you think will happen to the direction of the magnetic field if the direction of the current is reversed?

Let's repeat some of Oersted's observations to try to answer these questions. We'll study the magnetic field in a plane perpendicular to a long, straight current-carrying wire. To do this, set up the experiment as shown in Fig. 27.1, with the battery, switch, and wires all in series. The wire should run through a hole in the plexiglass sheet so that you can position a set of compasses near the wire. Check your compasses *before* starting the experiment to ensure the needles are all pointing north and swinging freely when the current is off.

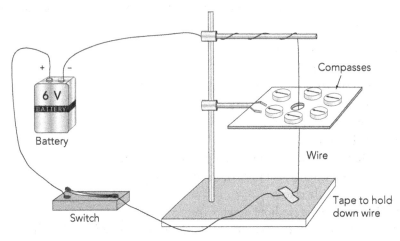

Fig. 27.1. Apparatus for repeating Oersted's observations on the magnetic field produced by a current. The switch is wired in a normally-open position, only making contact when you hold it down.

Important: To preserve the battery, turn on the current only when you are making observations. (The only resistance in the circuit is from the wire, so the current will be large.)

27.2.2. Activity: The Magnetic Field near a Wire

a. Wire the battery so that current passes *upward* along the wire through the plexiglass (from below to above). Momentarily close the switch and observe the behavior of the compasses. If you don't see anything, try moving the compasses closer to the wire. Explain what happens to the direction of the needles when the current is on. The diagram below shows an overhead view of the situation, with the wire (and current) coming up out of the paper. On the diagram, show the direction of the magnetic field at different points around the wire with arrows representing the orientation of the compass needles.

b. Next, reverse the leads to the battery so that the direction of the current is reversed (going from top to bottom). In the overhead view, this means the current is directed down into the paper. Once again, use the diagram below to show the direction of the magnetic field at different points around the wire.

c. You should see that the magnetic field circles around the wire and that the direction of the field depends on the current direction. Let's see if we can come up with a rule for determining the magnetic field direction. Hopefully, you remember the right-hand rule we used to describe the direction of angular velocity $\vec{\omega}$. Can you use a similar rule here to describe the direction of the magnetic field based on the direction of the current? Explain.

d. Try moving the compasses farther away from the wire. Does the strength of the magnetic field appear to change? If so, does it increase or decrease? Does this make sense?

e. Finally, disconnect the 6V lantern battery and replace it with a 1.5V D-cell battery to reduce the current in the circuit. Briefly close the switch. What happens to the apparent strength of the magnetic field when the current in the wire is decreased? Does this make sense?

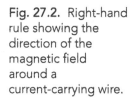

Fig. 27.2. Right-hand rule showing the direction of the magnetic field around a current-carrying wire.

You should have seen that the magnetic field circles around a current-carrying wire. The direction of the field can be described using a modified right-hand rule (see Fig. 27.2). Imagine grabbing the wire using your right hand with your thumb pointing in the direction of current flow; your fingers will then curl around the wire in the direction of the magnetic field. At any given point along the wire, there is a magnetic field curling in a circle around the wire. The strength of this field decreases if we move farther away from the wire or if we reduce the current.

What happens if we combine multiple current-carrying wires? Does the principle of superposition, which works when we combine electric fields, also work when combining magnetic fields? In the next activity we will examine two different configurations of wire to try to answer this question.

27.2.3. Activity: Magnetic Fields from Different Wiring Arrangements

a. How do you predict the strength of the magnetic field due to two wires carrying current in the *same* direction (Fig. 27.3a) will compare to the strength due to a single wire carrying the same current? Explain the reason for your prediction.

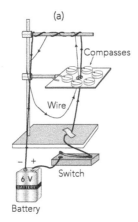

b. How do you predict the strength of the magnetic field due to two wires carrying current in *opposite* directions (Fig. 27.3b) will compare to the strength due to a single wire? Briefly explain.

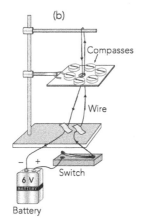

c. Adjust the wire in your set-up so that two lengths of wire run next to each other and carry current in the *same* direction (Fig 27.3a). The exact placement of the wires is not critical; what's important is that the wire runs through the hole in the plexiglass twice and that the currents in these two segments are in the *same* direction. Momentarily close the switch and observe the effect on the compasses. Can you tell if the magnetic field in this configuration is weaker, stronger, or the same as that from a single wire?

Fig. 27.3. Circuit with different orientations of wire sections. (a) Wires paired to carry current in the *same* direction; (b) Wires paired to carry current in *opposite* directions.

d. Next, adjust the wire so that two lengths of wire run next to each other but carry current in the *opposite* direction (Fig. 27.3b). Once again, the exact placement of the wires is not critical. Momentarily close the switch and observe the effect on the compasses. Can you tell if the magnetic field in this configuration is weaker, stronger, or the same as that from a single wire?

e. You should observe that the magnetic fields *superimpose*. With two currents running in the same direction, the total field is stronger (it turns out to be twice as strong, although this can't be quantified with our experiment). With two currents running in opposite directions, the total field is much weaker. Imagine you could get the two segments of wires carrying current in opposite directions in part (d) to be *exactly* on top of each other. Thinking about the principle of superposition, what should the net magnetic field be in the region around the wires? Briefly explain.

27.3 AMPÈRE'S LAW

The French physicist André Marie Ampère, excited by Oersted's observations of the magnetic behavior of current-carrying wires, immediately devoted a great deal of time to making careful observations of electromagnetic phenomena. These observations enabled Ampère to develop a mathematical expression relating the current in a wire to the resulting magnetic field produced. This equation is known as Ampère's law and is given by

$$\oint \vec{B} \cdot d\vec{s} = \mu_0 i_{\text{encl}} \quad \text{(Ampère's Law)} \tag{27.1}$$

where μ_0 is a physical constant called the magnetic *permeability* of free space, i_{encl} is the net electric current passing through the loop, and the circle on the integral symbol indicates that the integration takes place around a closed loop. Hopefully, this equation looks somewhat familiar, as it appears somewhat similar to Gauss's Law, given in Eq. (20.4):

$$\oint \vec{E} \cdot d\vec{A} = \frac{q_{\text{encl}}}{\varepsilon_0} \quad \text{(Gauss's Law)} \tag{27.2}$$

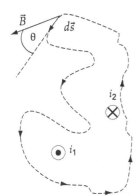

Fig. 27.4. An imaginary Ampèrian loop with net current $i_{\text{encl}} = i_1 - i_2$. The dotted line represents one of an infinite number of possible closed curves that enclose the two current-carrying wires. **Note:** Remember that the dotted loop represents an imaginary path, not a wire!

Both expressions involve closed integrals on the left side, which are proportional to either the enclosed charge or the enclosed current.

Conceptually, Ampère's law is a two-dimensional analog to Gauss's law and relates the line integral of the magnetic field around a closed curve to the current enclosed by that loop. The imaginary "Ampèrian loop," which can have any shape (as along as it doesn't cross itself), is broken up into an infinite number of little vectors $d\vec{s}$ lying tangent to the path (see Fig. 27.4). The dot product between $d\vec{s}$ and $\vec{B}$ picks out the component of the magnetic field that lies parallel to the curve: $\vec{B} \cdot d\vec{s} = B \, ds \cos\theta$. Each of these pieces is added up around a complete loop by the integral.

The enclosed current i_{encl} is simply the net current passing through the area enclosed by the loop. For the case shown in Fig. 27.4, there are two currents i_1 and i_2, with i_1 directed out of the page and i_2 into the page. The sign of the current is defined by the orientation of the Ampèrian loop and the right-hand rule. For the situation shown in Fig. 27.4, i_1 is positive, i_2 is negative, giving $i_{\text{encl}} = i_1 - i_2$.

As with Gauss's law, Ampère's law is always true but most useful for *symmetric* situations. In the next activity we will use Ampère's law to calculate the magnetic field produced by a current flowing through a long, straight wire. Hopefully, the steps will feel familiar based on our previous work with Gauss's law.

For the investigations in this section, you will need:

- 1 data-acquisition system
- 1 Hall-effect magnetic field sensor
- 1 insulated wire, 1 m long (16 AWG w/ thermoplastic insulation)
- 1 holder for wire coils (e.g., PVC pipe with groove for windings and cut-out for magnetic field sensor)
- 1 power supply in constant-current mode, 1A (or 1 D-cell battery, 1.5 V, with holder)
- 1 resistor, 2 Ω/2 W for limiting current (if using a battery)
- 1 ammeter, 1 A (if using a battery)
- 1 normally-open switch (if using a battery)
- 2 alligator clip leads, > 10 cm
- 1 Magnaprobe or small compass (optional)

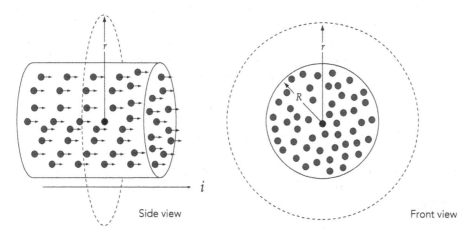

Fig. 27.5. An imaginary circular Ampèrian loop of radius r constructed outside a conductor of radius R carrying a current i. The moving charges are indicated by the small dots, which are coming out of the page in the front view.

Figure 27.5 shows two views of a wire carrying current i. In both views the black dots representing positive charge carriers flowing inside the wire (moving to the right in the side view and out of the page in the front view). We have drawn an imaginary Ampèrian loop that circles around the outside of the wire. Note that this shape is chosen based on the symmetry of the situation (you would not want to choose a square-shaped loop for this problem!).

27.3.1. Activity: The Magnetic Field Outside a Wire

a. Using your previous observations and the right-hand rule, sketch the direction of $\vec{B}$ along the imaginary loop in the *front-view* diagram.

b. Next, draw in a few vectors representing $d\vec{s}$ at different points on the imaginary loop, choosing them to point counter-clockwise around the loop (thereby defining the direction of positive current). What is the angle between $\vec{B}$ and $d\vec{s}$ at all point around the loop? Does it change, or is it always the same?

c. Does the magnitude of $\vec{B}$ change as you make your way around the loop? Why or why not?

d. Evaluate the integral on the left side of Ampère's law, using the fact that $\left|\vec{B}\right|$ is constant, the angle between $\vec{B}$ and $d\vec{s}$ is always zero, and the imaginary loop is a circle of radius r.

e. All that remains is to evaluate the right side of Ampère's law, which involves finding the net current enclosed by our imaginary loop. It's simple in this situation, as there is only a single wire and we are given the current! For this problem $i_{encl} = i$, the current in the single wire. Use this fact to solve for the magnitude of the magnetic field as a function of the distance from the wire.

You should have found that the magnitude of the magnetic field for a long, straight wire with current i is given by

$$B = \frac{\mu_0 i}{2\pi r} \quad \text{(outside a long, straight wire)} \tag{27.3}$$

where r is distance away from the wire. The direction of the field is specified by our new right-hand rule.

In the activity above, notice that for any position outside the wire ($r > R$), the radius of the wire doesn't matter (all the current is enclosed). If desired, we could also apply Ampère's law at a position *inside* the wire. Our imaginary Ampèrian loop would still be circle, but its radius would be $r < R$. In this case the full current i is not enclosed by the loop, so we would need to determine how much of the total current is enclosed by the loop (just like we did for a charge density with Gauss's law).

27.3.2. Activity: The Magnetic Field from Multiple Wires

Consider the two current-carrying wires shown in Fig. 27.6. Each wire carries 1.5 A of current but in opposite directions. We want to find the magnetic field at the three points labeled 1, 2, and 3.

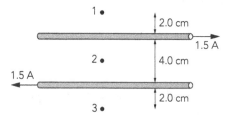

Fig. 27.6. Set-up involving two current-carrying wires.

a. Start with point 2. Use the result of Activity 27.3.1 to calculate the magnetic field at point 2 due to the top wire (both magnitude and direction). Then do the same for the magnetic field due to the bottom wire (once again at point 2). Then add these two fields, remembering that they are vectors and must be added vectorially!

b. Next, repeat this procedure to find the magnetic field at point 1.

c. Finally, repeat this procedure to find the magnetic field at point 3.
    Hint: Based on the symmetry of the situation, you may find your result to part (b) helpful.

The Magnetic Field at the Center of a Current Loop

In Activity 27.3.1 we derived an expression for the magnetic field outside a long, straight wire. One of the goals in this unit is to explore the effects of changing magnetic fields, and an easy way to produce a changing magnetic field is by varying the current in a wire. Typically, one uses a loop, or coil, of wire for this, and so we consider such an example in this section.

Single loop

Fig. 27.7. A single wire loop carrying a current i.

Multiple loops

Fig. 27.8. A multiturn loop carrying current i.

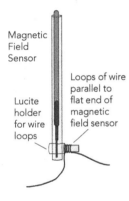

Magnetic Field Sensor

Lucite holder for wire loops

Loops of wire parallel to flat end of magnetic field sensor

Top view of Lucite holder

Fig. 27.9. Wire loops and tip of the magnetic field sensor. Note that the loop and the flat surface of the probe element of the sensor should be parallel. A small lucite holder can be used for making this arrangement.

27.3.3. Activity: The Magnetic Field in a Loop of Wire

a. On the basis of your observation of the magnetic field surrounding a straight wire, what *direction* do you think the magnetic field will be in the center of the single loop shown in Fig. 27.7? Think carefully about how you can use the right-hand rule from Fig. 27.2 for the case of a loop of wire.

b. How do you expect the magnitude of the field at the center of the loop to change if you make two loops? What about three loops such as in Fig. 27.8? Cite evidence from previous observations to support your prediction.

c. We now wish to measure the magnetic field in the center of a set of loops using a magnetic field sensor. To eliminate the effects of Earth's magnetic field, we will need to "zero" the sensor with the current turned off. Open up your data acquisition software and display a live reading of the magnetic field. Hold the magnetic field sensor out over your table (no wire loops involved). Slowly rotate the sensor—what do you see? Does the reading depend on the orientation of the sensor?

d. The sensor measures the magnetic field that is directed through the plane of the sensor. Because of this, you must orient the sensor so that the plane of sensor is parallel to the plane of the loop (and perpendicular to the direction of the magnetic field). Start by connecting the power supply and wire in series.[1] Then, wrap *one turn* of wire around the holder and carefully place the sensor so that it is near the center of the loop and measuring the magnetic field perpendicular to the plane of the loop.

With the current turned *off*, zero the magnetic field sensor. **Important:** Once you have zeroed the sensor, you cannot move the arrangement at all! Next, turn on the current to approximately 0.4 A (or briefly close the switch if you are using a battery). While the current is on, record the strength of the magnetic field due to the single loop of wire in the table that follows. Turn off the current when you are done, but be careful not to move the wire holder or sensor!

---

[1] If you are using a battery instead of a constant-current power supply, you will want to put a current-limiting resistor in series as well.

e. Next, wrap a second loop of wire around the holder. Turn on the current and measure the strength of the magnetic field. Record your measurements in the table. Repeat this process for up to four total loops. Remember, the wire holder and sensor cannot be moved during this process!

N (# of loops)	i (A)	$B_{measured}$ (mT)	$B_{per\ loop}$ (mT)
0	0	0	0
1			
2			
3			
4			

f. Using the data in the table, plot the strength of the measured magnetic field B versus the number of loops N. Print out or sketch your plot. Do your observations agree with your prediction in part (b)?

You should have seen that the magnetic field at the center of the coil is *proportional to the number of turns of wire* in the coil. In addition, we already know that the magnetic field is *proportional to the current* flowing through the wire. Putting these together we have

$$B \propto Ni$$

where N is the number of loops of wire and i is the current. Calculating the exact value of the magnetic field at the center of a circular coil of radius R is complicated and requires using the Biot-Savart law. Here, we simply quote the result:

$$B = \frac{\mu_0 Ni}{2R} \quad \text{(center of circular coil)} \tag{27.4}$$

where R is the radius of the coil.[2]

FARADAY'S LAW

27.4 MICHAEL FARADAY'S QUEST

In the nineteenth century the creation of a magnetic field by a current-carrying wire led investigators to consider a related question: is it possible for magnetism to cause a current to flow in a wire? Michael Faraday attempted to produce electricity from magnetism by reportedly putting a wire connected to a galvanometer near a strong magnet, but no current flowed. However, Faraday continued wrestling with this idea off and on for many years, and he eventually discovered

---

[2] This assumes the coil is tightly wound so that essentially all the turns are "on top of each other."

that he could produce a current in a coil of wire by using a *changing* magnetic field. This seemingly small feat ended up having a profound impact on civilization, as most of the electrical energy produced throughout history is a result of this process known as magnetic *induction*.

For the activities in this section, you'll need:

- 1 galvanometer (analog ammeter) showing both positive and negative current
- 4 assorted wire coils (with different areas and numbers of turns)
- 2 alligator clip leads
- 1 bar-shaped or rod-shaped magnet
- 1 U-shaped magnet

The goal of these observations is two-fold—first, to get a feel for what induction is like, and second, to discover what factors influence the amount of current induced in the coil. To start your observations, you should connect one of the coils to the galvanometer and fiddle around with the bar magnet in the vicinity of the coil.

27.4.1. Activity: Current from a Coil and Magnet

a. Play around with the coils and magnets and verify that you can induce a current in the galvanometer. Make a list of as many factors as possible that appear to determine the amount and/or direction of current that can be induced. **Note**: Your galvanometer may have a "push-to-read" button designed to protect it when not in use; if so, be sure to hold this button down while making your observations.

b. Can you find any way to generate a current in the coil if the magnet and coil are *not* moving?

You should be able to generate a current in the coil by moving the magnet *into* (or *out of*) the coil. Moreover, the amount and direction of current will depend on how fast you move the magnet, which pole is being used, the size and number of turns of the coil, etc. But it should be clear that a magnet *can* be used to produce an electric current! In fact, it is the magnetic *flux* through the coil that turns out to be important. As long as the magnetic flux through the coil is changing, there will be a current induced in the coil.

Before examining the factors that affect the induced current, we need to introduce a new term. A changing magnetic flux induces a current, and the electrical action that pushes the charges along the wire is called the *electromotive force*, abbreviated emf and denoted by ε. Because the induced current flows through wires that have some resistance, the electromotive force is equivalent to a potential difference along the wire. Therefore, it is the electromotive force (emf), which is measured in volts, that will appear in Faraday's law.[3]

27.4.2. Activity: Factors Affecting the Electromotive Force

a. How do you think the electromotive force (induced voltage) depends on the *rate* at which the magnetic flux changes through a coil? Feel free to try out the experiment again!

b. How do you think the electromotive force depends on the *number of turns* in the coil?

c. Based on parts (a) and (b), write down a trial equation that describes the induced emf as a function of the factors you think are important. For example, "the induced emf is proportional to … "

27.5   A MATHEMATICAL REPRESENTATION OF FARADAY'S LAW

Let's consider a magnetic field $\vec{B}$ that is uniform in space at any given moment but changes with time (getting stronger or weaker). How much electromotive force ε will be induced in a coil of wire? Quantitative experiments show that the electromotive force induced in a coil of wire depends on the rate of change of

---

[3] Electromotive force is the term used when a current is produced by a non-electrical source. For example, a battery uses chemical energy to produce a current and is therefore a source of emf. A 1.5 V battery supplies an electromotive force of $\varepsilon = 1.5$ V.

magnetic flux Φ_{mag}. Remember that flux is given by the dot product of the field and area vector, so the magnetic flux for a uniform magnetic field is[4]

$$\Phi_{mag} = \vec{B} \cdot \vec{A} = BA \cos\theta \qquad \left(\text{magnetic flux for uniform } \vec{B} \text{ field}\right) \qquad (27.5)$$

where $\vec{B}$ is the magnetic field, $\vec{A}$ is the area vector for the coil of wire, and θ is the angle between $\vec{B}$ and $\vec{A}$.

To form Faraday's law, we recognize that the rate of change is given mathematically by the derivative with respect to time, and we also include the dependence on the number of turns of wire (that you may have observed). Putting this all together gives us the following expression for the induced electromotive force (which we state without formal proof):

$$\varepsilon = -N\frac{d\Phi_{mag}}{dt} \qquad \left(\text{Faraday's Law for a coil with N turns}\right) \qquad (27.6)$$

There is one important part of this expression that deserves some discussion: the negative sign. In Activity 27.4.1 you probably noticed that the sign of the induced current (and hence the sign of the emf) depends on which pole is being moved near the coil. Careful experiments reveal that the appropriate sign needed in Faraday's law is negative. We will return to this issue in a later activity, but for now we take it as given.

For the activities in this section, you'll need the following equipment

- 1 solenoid with a 110 VAC plug
- 1 pickup coil with a small bulb attached (with a larger diameter than the solenoid)

An Example Using Faraday's Law

Let's consider a specific arrangement of magnetic field and coil in order to work through an example using Faraday's law. Figure 27.10 shows a coil oriented such that the plane of the coil is vertical and the vector $\vec{A}$ points to the right. We are free to choose our axes however we wish, and we arbitrarily pick the z-direction to be in the same direction as $\vec{A}$ so that $\vec{A} = A\,\hat{z}$. The magnetic field in the region of the coil can point in any direction, and we show it pointing up and to the right as compared to $\vec{A}$ such that the vectors $\vec{B}$ and $\vec{A}$ have an angle θ between them.

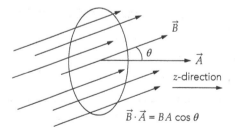

Fig. 27.10. Magnetic flux through an area A is the dot product of the magnetic field vector and the vector normal to the area.

---

[4] If the magnetic field were not spatially uniform, one would need to take the dot product of $\vec{B}$ and $d\vec{A}$ at each point and integrate over the area of the coil.

If the flux were always constant (in both magnitude and direction), its time derivative would be zero and there would be no induced emf. For this example we'll assume that $\vec{B}$ is spatially uniform and always oriented along the same axis, but that it changes in time in a sinusoidal manner $\vec{B}(t) = \vec{B}_0 \sin \omega t$, where $\vec{B}_0$ is a vector that points along the direction of the magnetic field and whose length corresponds to the maximum magnitude of the field, and ω is the (angular) frequency.

27.5.1. Activity: Applying Faraday's Law to a Wire Loop

a. Assuming the coil has N turns and radius R, calculate the magnetic flux $\Phi_{mag}(t)$ through the coil as a function of time.

b. Use this result and Eq. (27.6) to calculate the induced emf $\varepsilon(t)$ in the coil as a function of time.

c. You should have found that $\varepsilon(t) = -N\pi R^2 \omega B_{0z} \cos \omega t$, where $B_{0z} = B_0 \cos \theta$ is the z-component of $\vec{B}_0$ (resulting from the dot product with $\vec{A} = A\,\hat{z}$). The graph below plots $B_z(t) = B_{0z} \sin \omega t$ for two-and-a-half cycles of magnetic field oscillations. On the graph below, plot the shape of $\varepsilon(t)$ over the same time period (the vertical scale is arbitrary). Be careful to line up the two graphs properly!

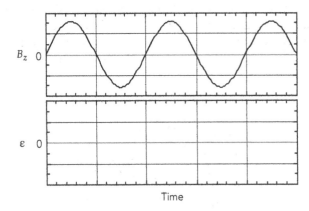

d. Suppose we change how the magnetic field varies in time so that instead of a sinusoidal curve, $B_z(t)$ varies in a triangular fashion, as shown in the following diagram. Just using this plot, sketch the shape of the induced $\varepsilon(t)$ on the graph below (the vertical axes are arbitrary, so no math needed!).

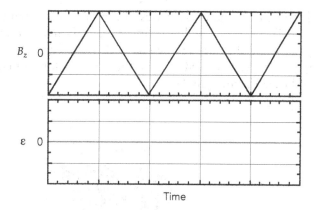

Mathematically, the induced emf comes from a time derivative of the magnetic flux through the coil. A time derivative can be seen graphically from the instantaneous slope of the appropriate graph. For the situation in the previous activity where only the strength of the magnetic field was changing, the induced emf is proportional to the negative slope of the $B_z(t)$ plot. Note that although Faraday's law allows us to calculate the induced emf, the resulting induced current must be calculated using Ohm's law.[5]

Summary of Faraday's Law

According to Faraday's law, there will be an induced emf any time the magnetic flux changes. Because the magnetic flux depends on the dot product of the magnetic field and the area vector of the coil, there are three independent ways the flux can change:

- the strength of the magnetic field through the coil can increase or decrease (e.g., a magnet is brought closer to the coil or the field varies in time)
- the area of the coil of wire can change (e.g., one side of a square loop could be a sliding contact that moves to increase or decrease the area)[6]
- the angle of the coil area vector relative to the magnetic field can change (e.g., the coil or the magnet can rotate about an axis)

A change in any one of these variables will induce an emf in the coil and, in principle, all three could change simultaneously.

Visualizing the Induced Current Using a Solenoid

Earlier, we saw a galvanometer needle deflect when pushing a magnet into a coil, but is an induced current significant enough to do something meaningful? To demonstrate that it is, we will use a device known as a *solenoid*. A solenoid

---

[5] Technically, most circuits will also have a property known as inductance, so finding the resulting induced current is a little more complicated. We will ignore that difficulty here.

[6] The area of the coil only matters if changing the area changes the magnetic flux. For example, it doesn't matter if the area changes if there is no magnetic field present in that region.

consists of a long wire wound into a series of circular loops, each one next to the previous one. This creates a hollow cylinder shape with one continuous coil of wire wrapped around the outside (see Fig. 27.11). If a current runs through the wire, Ampère's law can be used to show that there will be a magnetic field running down the axis of the cylinder.[7]

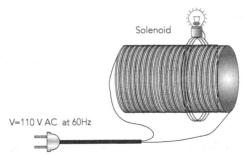

Fig. 27.11. A pickup coil with a light bulb attached is surrounding a solenoid (but not touching it).

In this demonstration, we will plug the two ends of the wire into an electrical socket in the wall. Common household wiring uses *alternating current* ("AC"), so the voltage as a function of time across the two leads of the socket varies sinusoidally.[8] The result is that the current in the solenoid will also vary sinusoidally, and hence the magnetic field generated by that current does as well. What happens when a coil of wire with a light bulb attached to it is placed over the solenoid as shown in Fig. 27.11? **Note**: The coil attached to the light bulb is not touching the solenoid in any way!

27.5.2. Activity: Induced Current Demonstration

a. What do you predict will happen when a coil with a bulb attached to it surrounds a solenoid with a changing magnetic field? Explain your response.

b. Now observe what happens. Was your prediction correct?

---

[7] In a "perfect" solenoid that is infinitely long, the field is directed entirely down the axis of the solenoid, and there is no field outside the cylinder.

[8] In the United States and Canada, the frequency of oscillation is 60 Hz with an RMS amplitude of 120 V.

27.6   LENZ'S LAW

We are now ready to return to the negative sign in Faraday's law, which tells us that the induced emf is opposite (the negative of) the rate of change of magnetic flux through the loop. But notice that it's not really clear what a "positive" (or "negative") emf represents. After all, we are dealing with a loop of wire that doesn't really have a beginning or an end! The easiest way to make sense of this is to think about the direction of the induced current, which can be found using something known as *Lenz's law*.

Lenz's law says that the direction of the induced current is such that the magnetic field created by the induced current *opposes* changes in the magnetic flux through the loop. Another way of saying this is that the system attempts to maintain the flux it currently has—if the flux is increasing the current flows in an attempt to counteract this increase, whereas if the flux is decreasing the current flows in an attempt to prevent this decrease.

Lenz's law is a bit confusing, so it's probably best discussed with an example. Before doing the example, your instructor may choose to show a brief demonstration (or two) that gets at the heart of Lenz's law. This can be accomplished with following equipment:

- 1 Lenz's Law Demonstrator (e.g., long aluminum tube along with both magnetic and non-magnetic cylinders that fit inside)
- 1 solenoid with a 110 VAC plug
- 1 (or more) metal rings that fit over the solenoid (aluminum, copper)

27.6.1. Activity: Faraday's Law Demonstrations

a. Imagine a non-magnetic, cylindrical object is dropped through a two-meter (non-magnetic) conducting tube. Roughly how long do you predict it will take to reach the bottom?

b. Now imagine you drop a cylindrical magnet through the same metal tube. Do you predict anything will be different? Why?

c. Test out both cases. What do you observe? Try to briefly explain, at least qualitatively, what you think might be happening (the next activity will examine this behavior in more detail).

d. Next, imagine you place a metal ring over a vertically-oriented solenoid (like the one from Activity 27.5.2). You now plug in the solenoid to the electrical socket in the wall so that there is a changing magnetic field down the axis. Predict what you think will happen, if anything, to the metal ring.

e. Observe what does happen. Is this what you predicted? Try to briefly explain, at least qualitatively, why you think this occurs.

Hopefully, these two demonstrations impressed upon you the dramatic effects that result from Faraday's (and Lenz's) law. In the next activity, we investigate Lenz's law more carefully to explain these phenomena.

Consider a set-up similar to that of Activity 27.4.1, where we pushed the north pole of a bar magnet into a loop of wire (see Fig. 27.12). Lenz's law says that the induced current in the loop tries to maintain a constant magnetic flux (it *opposes* the change in the flux). Now, the induced current will create its own magnetic field (called the "induced field"), and Lenz's law says the direction of this induced magnetic field is such that the original flux is maintained.

Fig. 27.12. A current meter showing an induced current as the north pole of a magnet is pushed toward the loop, thereby increasing the magnet flux through the loop.

27.6.2. Activity: Lenz's Law Example

a. Consider the situation in Fig. 27.12. When the magnet is far from the loop, explain why there is essentially no magnetic flux through the loop.

b. As the magnet gets closer to the loop, will there be any flux through the loop? In which direction is the magnetic field pointing through the loop?

c. If the system attempts to maintain the original flux (essentially zero when the magnet is far away), in what direction should the *induced* magnetic field point? **Reminder**: The induced magnetic field is a result of the induced current.

d. Based on your answer to part (c) determine the direction of the induced current in the loop required to produce the induced magnetic field. Draw in the direction of current on Fig. 27.12.

e. Think about the magnetic field of the magnet and the induced field created by the (induced) current. How will these fields interact? In particular, do the magnet and loop exert a force on each other? If so, in what direction is the force on the magnet? You may find it useful to think of the induced current in the loop as a magnetic dipole moment like we did in Section 26.7. Briefly explain.

You should have found that the induced current in the loop runs clockwise (when viewed from the left) in order to produce an induced magnetic field that points to the right. This induced field pointing to the right tends to cancel out the increasing left-pointing field from the bar magnet as it moves closer. From the perspective of magnetic flux, the bar magnet moving to the left changes the magnetic flux through the loop, and the induced magnetic field does its best to cancel out this change.

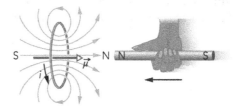

Fig. 27.13. The direction of induced current and induced magnetic field is shown when the north pole of a magnet is pushed to the left (toward the loop). Note how the induced field opposes the change in flux. As indicated, the loop with induced current can be thought of as a magnetic dipole, resulting in a repulsive force between the loop and magnet.

With the induced current in the loop, the loop now acts like a little disc magnet (or magnetic dipole) with the effective north pole of the current loop pointing to the right. As the magnet is pushed to the left, there are two north poles getting closer together. Since like poles of magnets repel, the bar magnet and current loop feel repulsive forces: the bar magnet feels a force to the right, while the current loop feels an equal and opposite force to the left. Importantly, this means that you must do *work* to push the magnet through the loop. It is this work that is ultimately responsible for the induced emf in the loop!

Let's return to the demonstration from Activity 27.6.1, where we dropped a magnet through a conducting (but non-magnetic) tube. We can use the results of Activity 27.6.2 to explain this phenomenon.

27.6.3. Activity: Magnet Falling in a Conducting Tube

a. In this demonstration we saw that the magnet fell through the conducting tube very slowly. Let's use Faraday's and Lenz's laws to explain why this happens. We start by considering the beginning of the process when the magnet is just about to enter the tube. Physically, where is the "loop" where the induced current flows? (After all, there is no wire in the traditional sense.)

b. Assuming the north pole of the magnet is pointing down, in what direction will the induced magnetic field from the tube point? Therefore, in what direction will the induced current in the tube flow?[9]

c. Explain why the induced current at the top of the tube causes the magnet to fall more slowly than usual.

d. Finally, explain why the magnet continues to fall slowly as it moves through the tube. Be sure to discuss where the additional "loops" are and what outside force is responsible for the process continuing.

---

[9] Such induced currents in a conductor due to changing magnetic flux are called "eddy currents."

VERIFICATION OF FARADAY'S LAW

27.7 VERIFYING FARADAY'S LAW

Our goal in this section is to perform a quantitative investigation of Faraday's law to experimentally verify Eq. (27.6):

$$\varepsilon = -N\frac{d\Phi_{\text{mag}}}{dt}$$

To do so, we must first *generate* the magnetic field required to produce the flux. While it's possible to use a permanent magnet, in practice it is easier to use another coil of wire with a current running through it. We will therefore have two separate coils of wire, each of which has many turns of wire to amplify the effect:

1. A large-diameter *field coil*, through which we run a current to produce a changing magnetic field.
2. A small-diameter *pick-up coil*, which is placed in the changing magnetic field so that an induced emf is generated.

A dual-trace oscilloscope can be used to display the voltage in both the field coil and the pickup coil, and we will try to verify Eq. (27.6) using this information. For the activities in this section, you will need the following equipment:

- 1 large-diameter field coil (200 turns works well)
- 2 different small-diameter pick-up coils (e.g., 400-turn and 2000-turn)
- 1 resistor, 1.2 kΩ
- 1 resistor, 10 kΩ
- 1 signal / function generator
- 1 two-channel oscilloscope
- 1 meter stick or ruler
- 1 protractor

A possible experimental setup is pictured in Fig. 27.14. A function generator creates a time-dependent voltage across the field coil, which in turn produces a

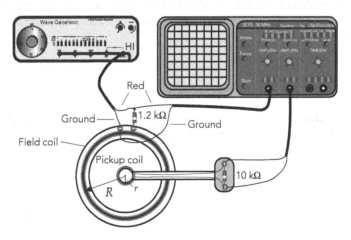

Fig. 27.14. Apparatus to test Faraday's law.

time-dependent current. This current then generates a changing magnetic field that is spatially uniform near the center of the field coil. The smaller pick-up coil is placed at the center of the field coil so that it experiences a changing magnetic flux, thereby generating an emf in the pick-up coil. An oscilloscope measures the induced emf in the pick-up coil, along with the voltage in the field coil for reference.

The Theory

The Field Coil: In Activity 27.3.3 we stated that the strength of the magnetic field at the center of a current-carrying coil is given by Eq. (27.4):

$$B_f(t) = \frac{\mu_0 N_f i_f(t)}{2R_f}$$

Note that we have added the subscript "f" to each variable to denote the field coil, and the time dependence of the current and magnetic field are explicitly shown. By measuring the voltage across the field coil and using Ohm's law, we can determine the current through the field coil (and hence the magnetic field in the region of the pick-up coil).

The Pick-up Coil: We now consider the pick-up coil in which the induced emf is given by Faraday's law

$$\varepsilon_p(t) = -N_p \frac{d\Phi_{\text{mag}}}{dt}$$

where we have added the subscript "p" for pick-up coil. Here, $\Phi_{\text{mag}} = \vec{B}_f \cdot \vec{A}_p$ is the magnetic flux through the pick-up coil, which depends on the magnetic field produced by the field coil and the area of the pick-up coil.[10]

27.7.1. Activity: Calculation of the Induced Voltage

a. Calculate the magnetic flux Φ_{mag} through a pick-up coil of radius R_p due to the magnetic field produced by the field coil. Put your answer in terms of the parameters of the field and pick-up coils, the angle between the two coils, and any physical constants. **Note**: Although Fig. 27.14 shows the planes of the coils parallel to each other, keep your expression general so that the two coils could be oriented at any angle with respect to each other.

b. Calculate the time derivative of this magnetic flux, assuming all quantities *except the current in the field coil* are constant.

---

[10] Once again, we assume the magnetic field produced by the field coil is spatially uniform over the region of the pick-up coil (a good assumption if the diameter of the pick-up coil is significantly smaller than that of the field coil).

c. Finally, determine the induced voltage in the pick-up coil in terms of all these quantities.

If you were careful with your calculations, you should have found that

$$\varepsilon_p(t) = -N_p \left(\frac{\mu_o N_f}{2R_f} \right) \left(\pi R_p^2 \right) \cos \theta \frac{di_f(t)}{dt} \tag{27.7}$$

where θ is the angle between $\vec{B}_f$ (from the field coil) and $\vec{A}_p$ (of the pick-up coil). Notice that the induced voltage in the *pick-up* coil depends on the time derivative of the current in the *field* coil.

The Experiment

As shown in Fig. 27.14, the field coil rests on the table with the pick-up coil at its center. A signal generator is connected to the field coil through a resistor, as well as to channel 1 of the oscilloscope; channel 1 thus shows the voltage as function of time in the *field* coil, $V_f(t)$. Channel 2 of the oscilloscope measures the voltage across a resistor connected to the *pick-up* coil, and thus shows the induced emf in the pick-up coil as a function of time, $\varepsilon_p(t)$.[11] By displaying both $V_f(t)$ and $\varepsilon_p(t)$, we can determine whether Eq. (27.7) (Faraday's law) provides an accurate description of this situation. Note that the oscilloscope should be "triggered" from channel 1 (the field coil signal).

If the experiment is not already set up, connect the signal generator and oscilloscope as shown in Fig. 27.14. We will start with a *triangle wave* voltage from the signal generator, in which the voltage applied to the field coil alternately increases and decreases linearly in time. This voltage produces a triangle-wave current in time in the field coil (see Fig. 27.15).

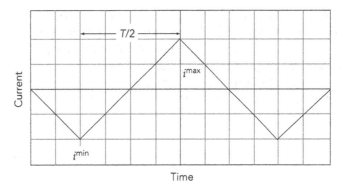

Fig. 27.15. A graph of current vs time in the field coil when using a triangle wave input from the function generator.

---

[11] The resistors are added to the field and pick-up coils to make it easier to measure the voltages across the coils.

27.7.2. Activity: Experimental Measurements and Comparison

a. Below, make an accurate plot the induced emf in the *pick-up* coil as a function of time. Be sure to include proper labels and units on your graph. What shape is the signal, and is this what you expect based on the graph in Fig. 27.15? Briefly explain.

b. Record the amplitude of the induced emf in the *pick-up* coil. **Remember**: The amplitude is the peak the signal reaches above (or below) zero. Your oscilloscope may have a "cursors" function to assist with this.

c. Next, let's determine values for the experimental parameters. Measure (or read off) the value of each parameter and record it below:

    i. Resistance of the field coil (the total resistance from one end to the other, *including* the attached resistor):[12] $Z_f =$

    ii. Number of turns in the field coil: $N_f =$

    iii. Number of turns in the pick-up coil: $N_p =$

    iv. Radius of the field coil (use the *average* radius): $R_f =$

    v. Radius of the pick-up coil (use the *average* radius): $R_p =$

    vi. Angle between the magnetic field $\vec{B}_f$ of the field coil and the area vector $\vec{A}_p$ of the pick-up coil: $\theta =$

d. To use Eq. (27.7) we need to determine $di_f(t)/dt$, the rate of change of current in the field coil. Channel 1 of the oscilloscope is measuring the *voltage* across a resistance for the field coil as a function of time, or $V_f(t)$. Start by making an accurate plot of the voltage in the *field* coil as a function of time. Be sure to include proper labels and units on your graph. Is the shape what you expected based on the field-coil current shown in Fig. 27.15?

---

[12] Because we are already using the variable R_f for the radius of the field coil, we need to choose a different variable to represent its resistance.

e. Using Ohm's law, make an accurate plot of the current in the *field* coil as a function of time. Be sure to include labels and units on the graph. **Note**: The shape will be as shown in Fig. 27.15; the point here is to determine the specific current and time values.

f. Use your result from part (e) to determine the rate of change of the current in the *field* coil during one of the sweeps. This is $di_f(t)/dt$. **Hint**: Since the change in current over any segment is linear, there is an easy way to get this derivative!

g. Finally, plug all the numbers into Eq. (27.7) to determine the *predicted* amplitude of the induced emf in the pick-up coil. How does the predicted value compare to your measured value from part (b)?

Hopefully, you found reasonable agreement between the theoretical prediction and the experimental measurement. To conclusively confirm Faraday's law, we would need to perform a series of measurements and carefully consider sources of uncertainty. Instead, we will simply assume Faraday's law holds and explore how the induced emf depends on some of the parameters in Eq. (27.7).

27.7.3. Activity: Dependence of Induced emf on Experimental Parameters

a. Increase the *frequency* of the field-coil triangle wave produced by the function generator by a factor of two. **Note**: You may need to adjust both the multiplier button and the frequency knobs on the wave generator (be sure the amplitude remains the same). Doubling the frequency means that the same voltage sweep occurs in *half* the amount of time. Adjust the time and voltage scales on the oscilloscope so that both voltage traces can be clearly seen on the screen. How does the amplitude of the *induced* emf change with this modification to the frequency of the oscillating magnetic field? Does it scale as you would expect based on Eq. (27.7)? Briefly explain.

b. Return the frequency (and amplitude) to its original value. Let's measure how the induced emf depends on the number of turns in the pick-up coil. Switch the pick-up coil to one with a different number of turns (if available). Adjust the time and voltage scales on the oscilloscope so that both voltage traces can be clearly seen on the screen. How does the amplitude of the *induced* emf change with this modification to the number of turns in the pick-up coil? Does it scale as you would expect based on Eq. (27.7)? Briefly explain.

c. Thus far we have kept the angle between the coils fixed at $\theta = 0°$. In the space below, *predict* what the induced emf graph will look like for angles of 90° (coils perpendicular) and 180° (pick-up coil flipped upside-down). Briefly explain.

d. Now measure the induced emf for $\theta = 90°$ and $\theta = 180°$. Do the results agree with your predictions?

e. Return the angle between the coils to $\theta = 0°$, and change the function generator to output a *sine wave* (like we considered in Activity 27.5.1). Look at the shape of the induced emf on the oscilloscope and sketch it in the space below. Is this what you expected? Explain why.

f. Finally, change the function generator to output a *square wave*. Look at the shape of the induced emf on the oscilloscope and sketch it in the space below. Are you able to explain why the graph has the shape it does?

27.8 MAXWELL'S EQUATIONS

The Scottish physicist James Clerk Maxwell was in his prime when Faraday retired from active teaching and research. He had more of a mathematical bent than Faraday and reformulated many of the basic equations describing electric and magnetic effects into a set of four (now famous) equations. These equations are shown below in their integral form for situations in which no dielectric or magnetic materials are present.

$$\text{Gauss's Law (electric fields):} \quad \oint \vec{E} \cdot d\vec{A} = \frac{q_{\text{encl}}}{\varepsilon_0}$$

$$\text{Gauss's Law (magnetic fields):} \quad \oint \vec{B} \cdot d\vec{A} = 0$$

$$\text{Maxwell-Faraday Law:} \quad \oint \vec{E} \cdot d\vec{s} = -\frac{d\Phi_{\text{mag}}}{dt}$$

$$\text{Ampère-Maxwell Law:} \quad \oint \vec{B} \cdot d\vec{s} = \mu_0 i_{\text{encl}} + \mu_0 \varepsilon_0 \frac{d\Phi_{\text{elec}}}{dt}$$

If we add the Lorentz force law for a single charge in an electric and magnetic field from Eq. (26.2)

$$\vec{F}^{\text{net}} = \vec{F}_{\text{elec}} + \vec{F}_{\text{mag}} = q\vec{E} + q\vec{v} \times \vec{B}$$

then this set of equations describes the complete content of all of classical electromagnetism.

Perhaps the most exciting intellectual outcome of Maxwell's equations is the prediction of electromagnetic waves ("light") and our eventual understanding of the self-propagating nature of these waves. This result was not fully appreciated until scientists abandoned the idea that all waves had to propagate through an elastic medium and accepted Einstein's theory of special relativity, which occurred in the early part of the twentieth century.

27.9 PROBLEM SOLVING

Electric generators are used in many types of power plants and other locations to generate electricity.[13] A basic design is shown in Fig. 27.16, where a coil of wire is placed in a region of uniform magnetic field as would be created by a large permanent magnet. The coil is then turned by applying a torque so that it rotates about its symmetry axis. This process creates an induced emf in the coil, which can be used to run a current through any "loads" attached to the generator. (Here the load is represented by a large resistor R, but the loads from large power plants are ultimately whatever we plug into the electrical outlets in our homes.) The sliding contacts allow the rotating ends of the loop to maintain electrical contact with the stationary wires attached to the resistor.

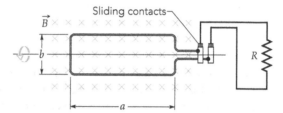

Fig. 27.16. Design of a simple electric generator. Although not shown, a torque must be applied (and work must be done) to rotate the coil.

For the activities in this section, you will need the following equipment:

- 1 mini-generator (hand-operated)
- 2 alligator clip leads (for the generator)
- 1 miniature light bulb
- 1 light bulb socket

27.9.1. Activity: Electric Generator

a. Assume the conducting coil has 1000 turns of wire with dimensions $a = 50$ cm and $b = 20$ cm. The uniform magnetic field is directed into the page with a magnitude of 0.5 T. If the coil is rotated at 1000 rpm (revolutions per minute), calculate the *maximum* value of the induced voltage

---

[13] Fossil fuel, biofuel, and nuclear power plants all operate in roughly this manner, burning different types of fuel to generate steam, which results in a torque that turns the coil(s). Wind turbines and hydroelectric dams also operate on this same principle, where either wind or water is used to physically turn the coil(s) of wire. Although some "solar fields" in the desert use the sun to boil water and function similarly to those above, the solar panels you seen on building roofs are photovoltaic cells that operate using a completely different principle, turning sunlight directly into electricity.

across the resistor. **Hint**: You will need to come up with an expression for how the angle between $\vec{B}$ and $\vec{A}$ changes in time.

b. You should find that the induced voltage in this process is not constant, but instead varies in time. Sketch a reasonably accurate plot of the induced voltage across the resistor as a function of time from $t = 0$ s to $t = 0.12$ s.

c. Even ignoring friction this process won't happen on its own—one must do *work* on the coil for it to turn. Use Lenz's law and/or the Lorentz force to briefly explain why conventional power plants must burn fuel in order to make the process above occur.

d. In Unit 16 we used a hand-operated generator to light up a bulb. This generator is nothing more than a miniature electrical powerplant, where instead of burning fuel to generate steam, your body does work to turn the crank (your body is "burning its own fuel"). Hook up the hand-operated generator to a light bulb and *gently* turn the crank of the generator to light up the bulb. Then remove the bulb from the circuit and once again *gently* turn the crank of the generator. What do you notice about turning the crank of the generator when the light bulb is attached as compared to when it is not? Briefly explain why this is the case.

INDEX

SYMBOLS USED IN THIS ACTIVITY GUIDE

$\vec{a}$	acceleration
A	area
$\vec{B}$	magnetic field
c	specific heat (per unit mass); speed of light
C_P	molar specific heat at constant pressure
C_V	molar specific heat at constant volume
C	capacitance
e	magnitude of charge of electron; base of natural logs, 2.71828 …
E	energy
$\vec{E}$	electric field
ε	emf (electromotive force)
ε_0	permittivity of free space
f	frequency; number of degrees of freedom
$\vec{f}$	frictional force (static or kinetic)
$\vec{F}$	force
g	local gravitational field strength
G	gravitational constant
i	current (assumed to be direction of positive charge carriers)
I	rotational inertia
$\vec{J}$	impulse
k_B	Boltzmann's constant
k	Coulomb's constant; spring constant
K	kinetic energy
L	latent heat
$\vec{\ell}, \vec{L}$	rotational momentum
m, M	mass
n	number of moles; number of density of charge carriers; number of turns per unit length
N	neutron number of nucleus; number of turns in a coil; total number of particles
$\vec{p}, \vec{P}$	momentum

P	power; pressure
q	charge of particle
Q	heat (thermal energy transfer); charge of system
$\vec{r}$	position vector
R	universal gas constant; resistance
S	entropy
t	time
T	temperature; period
U	potential energy
v	speed
$\vec{v}$	velocity
ΔV	electric potential difference
W	work
$\hat{x}, \hat{y}, \hat{z}$	unit vectors in the x, y, z directions
Z	atomic number of nucleus
α	angular acceleration (magnitude); alpha particle (He nucleus)
β	beta particle (electron)
γ	ratio of specific heats, C_P/C_V
θ	angle; rotational position
κ	dielectric constant
λ	wavelength; decay constant; charge per unit length
μ	coefficient of friction (static or kinetic); mass per unit length; permeability
ρ	mass (or charge) per unit volume; resistivity
σ	charge per unit area; conductivity
τ	torque (magnitude); time constant; mean time between collisions
$\vec{\tau}$	torque
Φ	flux of a vector field
ϕ	phase constant
ω	rotational speed; rotational frequency
$\vec{\omega}$	rotational velocity

PHYSICAL CONSTANTS

Local gravitational field strength (near Earth)	g	9.81 N/kg
Gravitational constant	G	6.67×10^{-11} N m^2/kg^2
Electron mass	m_e	9.11×10^{-31} kg
Proton mass	m_p	1.673×10^{-27} kg
Neutron mass	m_n	1.675×10^{-27} kg
Speed of light	c	3.00×10^8 m/s
Universal gas constant	R	8.31 J mol^{-1} K^{-1}
Boltzmann's constant	k_B	1.38×10^{-23} J/K
Avogadro's number	N_A	6.02×10^{23}/mol
Electric constant (permittivity)	ε_0	8.85×10^{-12} F/m
Coulomb constant	$k = \dfrac{1}{4\pi\varepsilon_0}$	8.99×10^9 N m^2/C^2
Elementary charge	e	1.60×10^{-19} C
Magnetic constant (permeability)	μ_0	$4\pi \times 10^{-7}$ T m/A
Unified atomic mass unit	u	1.66×10^{-27} kg

PHYSICAL PROPERTIES

Air (at room temperature and sea level atmospheric pressure)

Density	1.20 kg/m^3
Specific heat at constant pressure (C_p)	1.00×10^3 J kg^{-1} K^{-1}
Speed of sound	343 m/s

Water (at room temperature and sea level atmospheric pressure)

Density	1.00×10^3 kg/m^3
Specific heat	4.18×10^3 J kg^{-1} K^{-1}
Speed of sound	1.26×10^3 m/s

Earth

Density (mean)	5.49×10^3 kg/m^3
Radius (mean)	6.37×10^6 m
Mass	5.97×10^{24} kg
Atmospheric pressure (average sea level)	1.01×10^5 Pa
Mean Earth-moon distance	3.84×10^8 m

CONVERSIONS

Length

    1 in = 2.54 cm

    1 ft = 12 in = 0.3048 m

    1 m = 39.37 in = 3.281 ft

    1 yd = 3 ft = 0.9144 m

    1 km = 0.621 mi

    1 mi = 1.609 km = 5280 ft

    1 lightyear = 9.461×10^{15} m

    1 Å = 10^{-10} m

Area

    1 m^2 = 10^4 cm^2 = 10.76 ft^2

    1 ft^2 = 0.0929 m^2 = 144 in^2

    1 in^2 = 6.452 cm^2

Volume

    1 m^3 = 10^6 cm^3 = 35.3 ft^3

    1 ft^3 = 2.83×10^{-2} m^3 = 1728 in^3

    1 liter = 1000 cm^3 = 1.0576 qt = 0.0353 ft^3

    1 ft^3 = 7.481 gal = 28.32 liters

    1 gal = 3.786 liters = 231 in^3

Pressure

    1 Pa = 1 N/m^2 = 1.45×10^{-4} lb/in^2

    1 atm = 1.013×10^5 Pa = 14.7 lb/in^2 = 760 mm Hg

    1 bar = 10^5 Pa = 14.50 lb/in^2

Mass

    1000 kg = 1 t (metric ton)

    1 slug = 14.59 kg

    1 u = 1.66×10^{-27} kg = 931.5 MeV/c^2

Force

    1 N = 0.2248 lb

    1 lb = 4.448 N

Velocity

    1 m/s = 3.28 ft/s = 2.24 mi/hr

    1 mi/hr = 1.61 km/hr = 0.447 m/s = 1.47 ft/s

    1 mi/min = 60 mi/hr = 88 ft/s

Acceleration

    1 m/s^2 = 3.28 ft/s^2

    1 ft/s^2 = 0.3048 m/s^2 = 30.48 cm/s^2

Time

    1 day = 24 hr = 1.44×10^3 min = 8.64×10^4 s

    1 year = 365 days = 3.16×10^7 s

Energy

    1 J = 1 N m = 0.738 ft lb = 10^7 ergs

    1 cal = 4.186 J

    1 Btu = 252 cal = 1.054×10^3 J

    1 eV = 1.6×10^{-19} J

    1 kWh = 3.60×10^6 J

Power

    1 hp = 0.746 kW = 550 ft·lb/s

    1 W = 1 J/s = 0.738 ft·lb/s

    1 Btu/hr = 0.293 W

SOLAR SYSTEM

Body	Mean radius of orbit (m)	Mean radius of body (m)	Mass (kg)
Sun		6.96×10^8	1.99×10^{30}
Mercury	5.79×10^{10}	2.42×10^6	3.35×10^{23}
Venus	1.08×10^{11}	6.10×10^6	4.89×10^{24}
Earth	1.50×10^{11}	6.37×10^6	5.97×10^{24}
Mars	2.28×10^{11}	3.38×10^6	6.46×10^{23}
Jupiter	7.78×10^{11}	7.13×10^7	1.90×10^{27}
Saturn	1.43×10^{12}	6.04×10^7	5.69×10^{26}
Moon	3.84×10^8	1.74×10^6	7.35×10^{22}

VECTOR PRODUCTS

$$\vec{A} = A_x\hat{x} + A_y\hat{y} + A_z\hat{z}$$

$$\vec{B} = B_x\hat{x} + B_y\hat{y} + B_z\hat{z}$$

$$\vec{A} \cdot \vec{B} = A_xB_x + A_yB_y + A_zB_z$$

$$\vec{A} \times \vec{B} = \begin{vmatrix} \hat{x} & \hat{y} & \hat{z} \\ A_x & A_y & A_z \\ B_x & B_y & B_z \end{vmatrix} = \begin{matrix} (A_yB_z - A_zB_y)\hat{x} \\ -(A_xB_z - A_zB_x)\hat{y} \\ +(A_xB_y - A_yB_x)\hat{z} \end{matrix}$$